Between Artifacts and Texts

Historical Archaeology in Global Perspective

CONTRIBUTIONS TO GLOBAL HISTORICAL ARCHAEOLOGY

Series Editor:
Charles E. Orser, Jr., *Illinois State University, Normal, Illinois*

A HISTORICAL ARCHAEOLOGY OF THE MODERN WORLD
Charles E. Orser, Jr.

ARCHAEOLOGY AND THE CAPITALIST WORLD SYSTEM: A Study from
Russian America
Aron L. Crowell

BETWEEN ARTIFACTS AND TEXTS: Historical Archaeology in
Global Perspective
Anders Andrén

CULTURE CHANGE AND THE NEW TECHNOLOGY: An Archaeology of
the Early American Industrial Era
Paul A. Shackel

A Continuation Order Plan is available for this series. A continuation order will bring delivery of each new volume immediately upon publication. Volumes are billed only upon actual shipment. For further information please contact the publisher.

Between Artifacts and Texts

Historical Archaeology in Global Perspective

Anders Andrén

University of Lund
Lund, Sweden

Translated by

Alan Crozier

PLENUM PRESS • NEW YORK AND LONDON

Library of Congress Cataloging-in-Publication Data

Andrén, Anders, 1952–
     [Mellan ting och text. English]
     Between artifacts and texts : historical archaeology in global
perspective / Anders Andrén ; translated by Alan Crozier.
         p.   cm. -- (Contributions to global historical archaeology)
     Includes bibliographical references (p.   ) and index.
     ISBN 0-306-45556-0
     1. Archaeology and history.   I. Title.   II. Series.
CC77.H5A5313   1997
930.1--dc21   97-40651
                                                                     CIP

ISBN 0-306-45556-0

The original version of this volume was published in Swedish under the title *Mellan ting och text: En introduktion till de historiska arkeologierna*, © 1997 Brutus Östling Bokförlag Symposion, Stockholm and Stehag, Sweden.

© 1998 Plenum Press, New York
A Division of Plenum Publishing Corporation
233 Spring Street, New York, N.Y. 10013

Printed in the United States of America

Preface

This book is about historical archaeologies all over the world; about their history, their methods, and their raison d'être. The focus is on an existential question for archaeology: whether investigations of material culture are necessary at all when studying societies with writing. Is it not sufficient to read and interpret texts if we wish to understand and explain historical periods? This book has been written out of a conviction that archaeology *is* important, even in the study of literate societies. Yet the book has also been written out of a conviction that the importance of the historical archaeologies is not obvious to everyone. The disciplines have a tendency to be marginalized in relation both to history and to archaeology and anthropology, because the archaeological results are sometimes perceived as unnecessary confirmations of what is already known. Although I regard theoretical considerations as crucial for all scholarly work, I do not think that the solution to this marginalization can be found in any "definitive" theory that might raise the disciplines above the threatened tautology. Instead, I have found it more important to examine different methodological approaches in the historical archaeologies, to investigate how material culture and writing can and could be integrated. I am convinced that the tautological threat should be averted in the actual encounter of artifact and text. By problematizing this encounter, I believe that it is possible to create favorable methodological conditions for new perspectives on the past. It is only through such new and interesting results that the historical archaeologies can assert their importance.

The problems treated in the book have engaged me ever since I came into contact with archaeology 25 years ago. As a student, a digging and report-writing archaeologist, a museum worker, and a teacher and researcher, I have constantly touched upon questions concerning the relation between artifact and text. In retrospect it is easy to see that these questions did not just happen to come my way; on the contrary, I have been continually drawn to them. Despite this fascination with the intricate association between material culture and writing, however, a long time elapsed before the idea for this book finally took shape. When it did, it gave me the opportunity to combine two major interests—

archaeology and travel—in one and the same project. Since my own archaeological specialization—North European medieval archaeology—has its obvious geographic and chronological limits, I have tried as far as possible to obtain direct, physical experience of other places and areas treated by historical archaeology. I have therefore visited many of the areas discussed in the following pages, on journeys undertaken both before and after the idea for the book took shape.

When I began this project, there were relatively few studies that tried to go beyond the bounds of the individual historical archaeologies in quest of more general features. As my research has proceeded, however, several new works have been published with partially similar boundary-crossing perspectives. It seems as if the time is ripe for viewing the historical archaeologies as a unit. The new texts have in several cases been a source of help and inspiration, although I do not always share the other authors' perspectives and suggested solutions to the "crisis" of the historical archaeologies.

Acknowledgments

There are many people who deserve thanks in connection with this book. My gratitude is due to those who have helped me in various ways and made the book possible. Some people have assisted me in purely practical ways, others have given me inspiration and support at the right moment, and others have had a direct influence on the shape of the text by reading and discussing drafts. It all started in the spring of 1986, when I was invited to Uppsala by Thomas Lindkvist to lecture to the historical seminar on the relation between archaeology and history. One year later, thanks to Klavs Randsborg, I had a chance to work in Copenhagen, where I began for the first time to read in depth about the relation between archaeology and written sources. This resulted in a short article that in a way marks the introduction to this work. In the autumn of 1989 I was in Copenhagen again, following an invitation by Lotte Hedeager. I then had the privilege of working at the Center for Research in the Humanities, where Mogens Trolle Larsen, Michael Harbsmeier, and Michael Rowlands in particular kindled my interest in literacy as a problem, and made me think in more global terms. This was followed in the spring of 1990 by a sabbatical term in Cambridge, where I gradually began to pin down my problem in the University Library and the cold Haddon Library. During these months in Cambridge I benefited from important conversations, in particular, with Marie-Louise Stig Sørensen, Ian Hodder, J. D. Hill, and Gina Barnes.

The ideas from Cambridge had to lie dormant, however, because my regular teaching post did not give me enough time to continue. It was not until the spring of 1993, when I had the privilege of becoming a reader in historical archaeology, that I found time to write the book and conclude the enterprise. With a few interruptions, the Institute of Archaeology in Lund has been the base for my work. There are many people who deserve thanks for making this workplace an inspiring environment. In this context, however, I must above all mention two persons in the department: Hans Andersson and Jes Wienberg. Hans has constantly encouraged, read, and commented on my explorations in the borderland of medieval archaeology, and he has always handled with stoical calm the practical problems entailed by my absences. With

Jes I have carried on an incessant and inspiring dialogue about archaeology and historical archaeology for many years. His comments on different versions of my book have, as always, been critical and thought provoking.

Thanks to generous grants from the Faculty of Arts, the Elisabeth Rausing Memorial Foundation, the New Society of Letters in Lund, and the Swedish Institute, I have been able to undertake many trips abroad, and I have also been able to invite guest scholars from Europe and the United States to lecture about problems in the historical archaeologies. These guest lectures have provided important inspiration for this text. Other decisive stages in my work have been series of seminars on historical archaeology that I have held in Lund, Tromsø, and Stockholm. The seminars forced me to formulate certain problems more precisely, and the discussions led me to narrow down several crucial questions.

The present text was mainly written in Lund in the autumn of 1994 and in Athens in the spring and summer of 1995. The greater part of 1996 was devoted to the revision of the Swedish manuscript and the translation into English, which was financed by the Swedish Council for Research in the Humanities and Social Sciences. In connection with the translation, I have had rewarding discussions with Alan Crozier about the difficult art of recasting Swedish expressions in English forms. Ann Tobin has helped me with bibliographical data and has compiled the index. Tina Borstam has drawn some of the illustrations, whereas Bengt Almgren has made most of the figures camera-ready. In the hectic final phase, Gunnar Broberg read the manuscript and made significant comments from the point of view of the history of ideas. Special thanks are also due to Charles Orser, who has been kind enough to accept the book for the Contributions to Global Historical Archaeology series and who has made important comments on the text.

Finally, I wish to turn to my family. My parents have, as always, supported my work. Besides reading and commenting on the text, they have also been of invaluable assistance in their role of grandparents. My wife, Sanne, has shared her life with me for roughly as long as the idea of this book has existed. I am especially grateful for her constant encouragement and for her professional criticism during these years. It is also due to her, and her work in Greece, that the most intensive writing period took place during some dreadfully hot summer months in Athens. Every morning and afternoon during this summer on the way to and from our children's day-care center, I passed over the conduit through which the River Ilissos is channeled. In other words, every day during this hot summer I passed the place where Socrates, according to Plato's *Phaedrus*, discussed writing as a problem.

Contents

Chapter 1. The Paradox of the Historical Archaeologies 1

Chapter 2. Historical Archaeologies in Europe 9

The European Traditions 9
  The Past as Utopia: Classical Archaeology 10
  The Christian Golden Age: Medieval Archaeology 25

Chapter 3. Historical Archaeologies in the Middle East and
            Asia 37

The Middle East 37
  The Origin or Cul-de-Sac of History: Egyptology 38
  Paradise Lost: Mesopotamian Archaeology 43
  Defending the Faith: Biblical Archaeology 49
Asia 54
  Exotic and Familiar: Indian Archaeology 54
  History as Morality: Chinese Archaeology 62
  In Search of a New Identity: Japanese Archaeology 68

Chapter 4. Historical Archaeologies in Africa and America .. 73

Africa 73
  Heart of Darkness: African Archaeology 73
America 82
  America's Greece: Mexican Archaeology 83
  The Past as a Laboratory: Peruvian Archaeology 91
  The New Europe: Historical Archaeology in the United
    States 95

Chapter 5.   The Field of Historical Archaeology   105

Introduction   105
The Field as Transgressing Traditions   107
    Staging the Past: The Aesthetic Tradition   107
    Giving Meaning to Writing: The Philological Tradition   113
    Extending the Text: The Historical Tradition   120
    Writing the History of Artifacts: The Tradition of Cultural
       History   126
    Searching for Analogies: The Archaeological Tradition   131
    Summing Up   134
The Field as Modern Discourse   135

Chapter 6.   The Dialogue of Historical Archaeology   145

Introduction   145
Defining Practice   146
The Construction of the Context   153
    Correspondence   157
    Association   168
    Contrast   171
    Smaller and Larger Contexts   175

Chapter 7.   Conclusion: Historical Archaeology as
              a Methodological Perspective   179

References   185

Index   209

Between Artifacts and Texts

Historical Archaeology in Global Perspective

The Paradox of the Historical Archaeologies | 1

A text that is nothing other than an artifact, an artifact that is nothing other than a text has remarkably little to say.

—JOSIAH OBER, 1995:122

Archaeology is limitless. Archaeologists can study the first human beings in East Africa with the same interest as yesterday's kitchen garbage in Tucson, Arizona. Yet archaeology is also full of limits. Archaeology is not a coherent tradition covering the whole of human history, but rather a scientific field crossed by different traditions and separated by diffuse boundaries from other fields of scholarship.

An important boundary in the field of archaeology runs between disciplines studying periods without writing and those studying periods with writing. This boundary separates "prehistoric" archaeology from "historical" archaeologies, such as Egyptology, classical archaeology, medieval archaeology, and historical archaeology in the United States. The boundary, which is based on the presence or absence of writing, is a legacy of the breakthrough of modern human science in the middle of the nineteenth century. In the antiquarian tradition of the seventeenth and eighteenth centuries the boundary did not exist, since it was an impossibility. Antiquarian study was based on the idea that human history as a whole could be followed through texts; even Creation itself was known through the book of Genesis. The idea of a prehistory, with endless spaces of time beyond the horizon of writing, was a radical breach with the antiquarian way of thinking, and it was the very foundation of archaeology as a modern science. Yet at the same time when "prehistoric" archaeology was established as a modern science, classical archaeology was also created as a professional "historical" archaeology. The division into prehistoric and historic archaeologies can thus be traced back to the beginnings of modern archaeology.

The division of modern archaeology into subjects focusing on either literate or illiterate societies is also reflected in the duality of the concept of archaeology. On one hand, it can be perceived as a discipline concerned with the distant past, before the oldest texts. On the other hand, it can also be perceived as a research field focusing on material culture in all ages, regardless of whether texts exist. In principle, most archaeologists today perceive the concept in its limitless sense, but in practice the attitude is often different. The cleavage is particularly clear in the view of the historical archaeologies. Considerable work is being done in the historical archaeologies; in Scandinavia and Britain alone, more than half of all archaeological literature is devoted to historical periods. At the same time, the historical archaeologies play a paradoxically modest role in general archaeological surveys. There are often historiographic overview of individual branches of historical archaeology, but the general surveys are dominated by a "prehistoric" perspective. The debate in prehistoric archaeology is automatically placed in the center, whereas historical archaeologies are dismissed as a "Balkanization" of the subject (Trigger, 1989:12), or viewed as rather irrelevant (Hodder, 1991), or passed over in complete silence (Klindt-Jensen, 1975). That is why we can read about C. J. Thomsen's three-age system, but not that his famous "Guide" also encompassed the Middle Ages, or that his contemporary Thomas Rickman worked with the same typological and stratigraphic methods in his chronological studies of the English Gothic. That is why we can read about Gordon Willey's epoch-making surveys in Peru, but not about contemporary landscape inventories conducted by British classical archaeologists in Etruria. And that is why we can read about the emphasis on the symbolic value of artifacts in postprocessual archaeology, but not about the long tradition of symbolic interpretation in many of the historical archaeologies. Several of the surveys have an implicit assumption that "archaeological" thinking has mainly taken place in the prehistoric archaeologies and that little is to be gained from the historical archaeologies, even if a great deal of work is done in them.

It is true that the discussion in the historical archaeologies is divided and often difficult to survey, since it is often geared to specific historical periods or areas, and is rarely formulated in relation to the general archaeological debate. Like K. R. Dark (1995:196), however, I believe that the stereotyped view of the historical archaeologies needs to be changed. Debates are carried on in the historical branches of archaeology too, and they are not just pale copies of the debates conducted in the prehistoric archaeologies. That is why the prehistoric perspective in many historical surveys of archaeology is too one-dimensional and oversimplified. The archaeological tradition is reduced, and hence also the image of archaeology, since historiography actively helps to create an archaeological identity. If the limitless sense of archaeology is to be maintained, the history of research must broaden its perspective to include all topics concerned with material culture. Historiographic surveys should therefore also comprise the historical archaeologies as well, and bear in mind that archaeology borders not just on anthropology and history but also on aesthetics, philology, European ethnology, American folk studies, and religious studies. The history of archaeology, and hence the archaeological identity, can thus acquire more facets, if both the internal and the external definitions of archaeology are made less unequivocal.

A "total" history of archaeology along these lines remains to be written; here I shall only contribute fragments from the often neglected field of the historical archaeologies. Since the area is defined on the basis of writing, the relation between artifact and text will be in focus throughout. To give as multifaceted a picture as possible of the historical archaeologies, I apply both historical and thematic perspectives. It is essential to emphasize the historical lines, since they are often missing from today's debate, and since they can make the problems concerning artifact and text more obvious. Yet it is also important to stress the thematic perspectives to cross the often sharp disciplinary lines within the field and hence create the conditions for a more general debate. My desire to transgress disciplinary boundaries follows a tendency in recent years to see the historical archaeologies as a coherent field, with certain specific problems (e.g., Kardulias, 1994; Little, 1992; Morris, 1994; Schmidt, 1983; Small, 1995; von Falkenhausen, 1993).

As one point of departure for the study, I want to take the paradoxically contradictory view that exists of the historical archaeologies. On one hand, the presence of written sources is seen as a great advantage, since archaeology is always dependent on analogies in order to translate material culture into texts. Many people working in the prehistoric archaeologies look favorably, almost longingly, at "text-aided" archaeology (e.g., Clarke, 1971). On the other hand, the presence of texts can be seen as a great disadvantage, since it seems to leave little scope for archaeology by hampering the potential of archaeological analyses and interpretations. There is a constantly overhanging risk of tautology, and all historical archaeology can become, as Peter Sawyer's drastic criticism runs, "an expensive way of telling us what we know already" (cited from Rahtz, 1983:15). Some archaeologists claim that it is precisely the presence of written sources that has led to the characteristic "theorylessness" of many historical archaeologies, since writing appears to take on the same explanatory value as theories in "prehistoric" periods (e.g., Austin, 1990; Ellis, 1983). To put it in extreme terms, the paradox is that those who lack texts want them and those who have texts would like to avoid them. The historical archaeologies are thus characterized by tensions between necessary analogies and inhibiting interpretations.

My aim is to try to chart these tensions, above all by stressing the relation between material culture and another discourse, namely, text. Writing is a "technologizing" of the spoken word (Ong, 1982), which means that it records a partly different "version" of the past from the one preserved in material culture. My study thus focuses on the way in which two partially different representations of the past can be related to each other. I shall not, however, consider pictures, another important and much older representation system than texts. If I had included images as an element running right through my study, the section of the archaeological field would have looked different and would have encompassed all archaeologies concerned with periods from the Late Paleolithic onwards.

Looking at artifact and text as partly different discourses means that some form of fundamental distinction is maintained between the two media. Consequently, I believe that we cannot solve the problems in historical archaeology by abolishing the distinction. To claim in a literal sense that artifact is identical to text (cf. Christophersen, 1992; Wienberg, 1988) or to perceive artifact and text as equivalent semiotic signs (see Sonesson, 1992:299 ff.) only makes the problems less visible, but they do not disappear.

My emphasis on the relation between artifact and text means, moreover, that I will be examining the shared features in the historical archaeologies. I will not, therefore, look in any depth at the way the different archaeological specialities can be related to theoretical models for individual periods—for example, medieval archaeology in relation to the concept of feudalism (cf. Andrén, 1985:66 f.; Klackenberg, 1986), or American historical archaeology in relation to the concepts of capitalism and modernity (cf. Leone, 1988; Little, 1994; Orser, 1988b, 1996:57 ff.; Paynter, 1988). Instead, my focus on the relation between material culture and writing means that the field of historical archaeology is viewed as a special methodological perspective. In this context, however, I see "method" as something more than a pure technicality. The historiographic outline makes it possible to detect changes in this methodological attitude, and "method" hence becomes a critical awareness of the changeable nature of practice.

Since the focus here is on the relation between material culture and text, the occurrence of writing is the primary limiting factor in the study (Figure 1). Writing was a conceptual revolution that made it

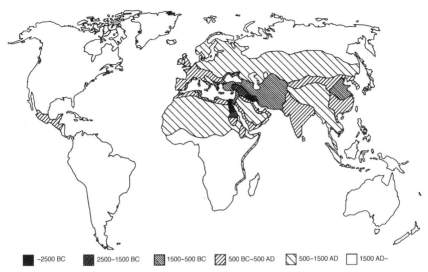

Figure 1. The distribution of deciphered scripts in time and place. The map is based on data on the occurrence of writing in different regions (Coulmans, 1996; Diringer, 1962, 1968; Djait, 1981; Hrbek, 1981; Marcus, 1992). The figure shows a minimal version of the chronological and geographic framework of historical archaeology. If ethnographic information, mythological narratives, and oral tradition were also included, this global view of historical archaeology could be modified in certain areas.

possible to render the spoken word in signs. This representation system, which is found in a multitude of different forms, has been spread over the world, probably from three different areas: Mesopotamia, China, and Central America (Diringer, 1962, 1968). Yet the presence of writing merely indicates the youngest possible limit for the historical archaeologies. Many cultures without texts of their own are known to varying extents from descriptions by outsiders, composed in areas with writing. In addition, oral tradition—whether recorded as written text or still living—can also go back beyond the oldest written evidence about an area (see Schmidt, 1978). The relevance of oral tradition for archaeology varies from region to region, however, and its chronological scope is often an unresolved controversy. A good example of the difficult problem of oral tradition is the renewed archaeological interest in the Indo-European languages (see Figure 30), since language per se must be viewed as the most fundamental form of oral tradition. The very idea of a common "Indo-European" origin for the related languages may be rejected, but if the idea is accepted, its importance for archaeology can vary significantly. Depending on the interpretative perspective, Indo-

European languages can be compared with material culture starting from the Early Neolithic, the Middle Neolithic, or the Early Bronze Age (cf. Mallory, 1989; Renfrew, 1987). Both ethnographic texts and oral traditions are important dimensions in the historical archaeologies, but because of the often unclear spatial and chronological scope of these types of tradition, the following survey primarily comprises the branches of archaeology that concentrate on areas and periods with writing. In these text-based disciplines, moreover, methodological issues are more explicit and hence discussed more often.

In addition, the question of writing and oral tradition concerns the very concept of "historical archaeology," and I shall briefly comment on this too. The concept can be used in two different senses: to designate archaeologies working with the modern era since around 1500 (see Orser, 1996:23 ff.; Orser and Fagan, 1995:6 ff.), or to define archaeologies focusing on all societies with writing over the last five thousand years or more. It is in the latter sense that I use the term, although it is not an entirely happy one, since it includes an ethnocentric hint that people without writing lack history (see Schmidt, 1983:64). Unfortunately, it is difficult to find a better alternative. The concept of "text-aided archaeology" (cf. Clarke, 1971; Little, 1992) has a one-sided bias toward archaeology, ignoring the fact that not only do texts aid archaeology, but the reverse is also the case. At present a term like "textual archaeology" evokes too obvious an association with postprocessual archaeology. Instead of a neologism, such as "grapho-archaeology," I have decided, after some hesitation, to retain the conventional terms "historical archaeology" and "historical archaeologies." These can be partly defended on the grounds that the concept of "history" is ambiguous. It stands not only for the history of humanity as a whole, but also in a limited sense for the mainly text-based discipline of history and for the potential to tell a story. For Hayden White (1987:55), the presence of writing means that "historical narrative" takes on a completely different character than in contexts where there are no texts. He therefore finds it justified to speak of "prehistory" as opposed to "nonhistory."

I will conclude this chapter by giving a short introduction to the following three chapters, in which the historical archaeologies are presented in a global outline. The aim of these chapters is to present briefly a number of disciplines in the field of historical archaeology to achieve a united starting point for the more thematic surveys. Although the historical archaeologies cover only parts of the archaeological field, the area is enormous. It is thus neither possible nor even desirable to present a total picture. I have deliberately concentrated on the European subject areas, since archaeology, as a "modern project," has its origin in Europe. At the same time, representative traditions from most other continents will be touched upon. I have primarily focused the introductory surveys on subject areas that are of interest in the present context, that is to say, disciplines that I have found interesting in relation to issues of artifact and text.

Since historiographic surveys of the individual disciplines are normally available, I do not seek to paint a complete picture of the history of research in each of the subjects. This global sketch should rather be seen as an attempt to capture essential perspectives in the different subject traditions. This concerns the general view of the areas and periods under study, and the methodologically important issue of the relation between material culture and writing. It goes without saying that I cannot have an equally profound knowledge of all the different subjects. However, experiences from my own subject, North European medieval archaeology, have guided me in my search for similarities and differences in the other disciplines. I have found good points of departure for my quest in surveys of research history, handbooks, anthologies, and debate articles. Any archaeological investigation in a historical period could potentially contribute to the survey, but I have included only those works that are brought up via the other texts.

Neither archaeology as a whole nor its special branches are God-given categories; instead, the divisions between different archaeological specialities are historical constructions. I have nevertheless chosen to present the following subjects as distinct areas, since the disciplines are often "self-defining"; that is, they can be limited by means of special designations, by patterns in modes of reference, and by more or less clear subject identities borne by the practitioners of the individual branches. I have taken particular interest in the professionalization of the various disciplines, since this shows when a sphere of knowledge is demarcated and acquires its characteristic profile. A more general interest in the past becomes a discourse that makes certain perspectives possible and others impossible. The final professionalization is normally expressed by the emphasis of a special identity in fieldwork, teaching, research, organizations, conferences, and journals. I have also devoted some attention to the preliminary stages of professionalization, which sometimes consist of a long professional "prehistory." Moreover, I have considered the interest shown in recent decades in the issue of artifact and text.

The perspectives on research history in the following summaries of the subjects are both intradisciplinary and extradisciplinary (see Liedman, 1978). Since there are normally historiographic surveys of most disciplines, I have primarily concentrated the intradisciplinary perspective on questions of artifact versus text. The extradisciplinary perspective means that I link up with the critical historiography in archaeology in the last two decades, when the political and ideological functions of the disciplines have been stressed (e.g., Bandaranayke, 1978; Keller, 1978; Mahler et al., 1983; Oyuela-Caycedo, 1994; Silberman, 1982, 1989; Trigger, 1984, 1989). This critical perspective is hinted at in various parts of the survey, but I deepen it only when the perspective can shed light on the "methodological" problems shared by the historical archaeologies.

Historical Archaeologies in Europe | 2

THE EUROPEAN TRADITIONS

The historical archaeologies in Europe are like a mosaic of different, partly overlapping traditions. The different parts consist of classical archaeology, provincial Roman archaeology, Byzantine archaeology, medieval archaeology, postmedieval archaeology, and industrial archaeology. In addition, historical archaeology can be found in some marine archaeology and in some "prehistoric" archaeology. In the latter case it is a question of protohistorical periods, such as the pre-Roman Iron Age in Central and Western Europe, and the post-Roman Iron Age in Northern and Eastern Europe. Taken together, the subjects cover all "historical" periods in Europe, but they do not represent a uniform archaeological tradition. The division into special branches, each with its own distinctive character, is largely due to the fact that the definitive "archaeological" professionalization took place at widely different times.

Despite the obvious division, there are nevertheless certain shared features. All the subjects have an indigenous European origin. They often have a long "prehistory," before the final professionalization, since material remains from the historical periods in Europe have been studied since the Renaissance, and they all concern fundamental questions of national and European identity.

In this context I have chosen to sum up only the two European archaeological traditions that I know best, namely, classical archaeology and medieval archaeology. Many of the characteristic features in these two disciplines can be found in the other subjects as well. For instance, the debate about artifact and text in classical archaeology and medieval archaeology is very similar to corresponding discussions in the other disciplines (cf. e.g., Alkemade, 1991; Champion, 1985; Gaimster, 1994; Harnow, 1992; Hill, 1993; Rautman, 1990; Scott, 1993).

The Past as Utopia: Classical Archaeology

Of the historical archaeologies in Europe, classical archaeology distinguishes itself most clearly, by virtue of its long prehistory and its early professionalization. The concept of "classical" in classical archaeology stands for "exemplary," which is a good reflection of the background and emergence of the subject. The very basis of the subject was that classical antiquity was studied to provide aesthetic and moral models for the present. This quest for a Utopia in the past was a general Western concern, which may explain why the subject is represented even in countries without a classical antiquity of their own, such as Scandinavia, Central Europe, the United States, and Australia.

It was only in Greece and Italy that classical archaeology came to play an important nationalist role in conjunction with the creation of the modern nation-states. The role of archaeology is particularly clear in connection with modern Greece. A knowledge of antiquity became a way for the great European powers to gain influence in the disintegrating Ottoman Empire, and it was simultaneously a significant driving force for Greek intellectuals to define the borders of modern Greece in the multicultural Balkans (cf. d'Agostino, 1991; Kotsakis, 1991; Morris, 1994).

The exemplary character of classical archaeology can also explain why its research history differs from that of provincial Roman archaeology (Niemeyer, 1968). The art and architecture of the Roman provinces was not viewed as exemplary. The artifacts were not exhibited in fine arts museums, and the Roman remains were mainly studied from the perspective of local history. When provincial Roman archaeology was professionalized—long after classical archaeology—the interest was chiefly focused on history and military history. Provincial Roman archaeology thus became a way to study Roman imperialism, creating historical clear associations with the imperialist policies of the great European powers (Hingley, 1991; Löfting, 1993; Potter, 1987).

The interest in antiquity never completely vanished in medieval Europe. The Carolingian Renaissance that started around 800 was a period when classical texts were collected and studied in detail, and ancient artifacts were actively reused (Braunfels and Schnitzler, 1965). An example of the significance of classical artifacts is the equestrian statue of Theodoric the Great, which Charles the Great had moved from Ravenna to Aix-la-Chapelle after he was crowned emperor in Rome. The removal of the statue, like the coronation, shows how Charles the Great deliberately invoked classical antiquity to define and configure his own political role. Several similar "renaissances" can be discerned in medieval Europe, but the direct preconditions for classical archaeology cannot be found until the Italian Renaissance in the fifteenth century. That was when an intellectual aristocracy in the city-states of northern Italy—especially in Florence—studied and imitated Roman antiquity as a means to renew aesthetic and moral attitudes in an increasingly secularized world. On the basis of detailed studies of classical sculptures and ruins, painters and architects in the early fifteenth century created the new style that posterity would call the "Renaissance" (see Figure 27). At the same time, humanists, through the study of ancient texts, searched for sources of human "refinement," for example, in the fields of rhetoric and moral philosophy. Several of these humanists also wrote historical works in which they began in earnest to use artifacts, monuments, and inscriptions both to complement and to check the veracity of classical texts (Weiss, 1969).

This aesthetic and historical interest in antiquity spread over Europe in the subsequent centuries, as the very basis for the antiquarian tradition (Schnapp, 1993). For a long time the Roman Empire continued to be the focal point of classical study, since it was more accessible linguistically, geographically, and politically than the remains of classical Greece. It was not until the mideighteenth century that an important turning point in the study of antiquity came, which was decisive for the direction of classical archaeology. This was when the modern concept of art was created, when pictorial art became accessible through public "salons" or exhibitions, and when the aesthetic experience began to be perceived as an end in itself (Emt and Hermerén, 1990). It was then that ancient Greece began to be the great aesthetic, moral, and political example for the emerging middle class in England, France, and the German states (see Vickers, 1990). The quest for Greece became a quest for democracy, philosophy, and beauty, and the material expression of this was neoclassicism. This style, with its emphasis on "noble simplicity and calm greatness," had a huge impact on architecture, fine arts, and decorative arts in Europe and North America (Honour, 1968).

Even though ancient Greece became the central object of antiquarian interest in the mideighteenth century, Italy retained something of its attraction. The interest was focused above all on the "intermediaries" and "heirs" of Greek art, namely, the Etruscans and early imperial Rome. Important sources of inspiration came from the excavations of the Roman cities of Herculaneum and Pompeii, which began in 1738 and 1748, respectively (Schiering, 1969:49 ff.). Classical archaeology thus acquired its characteristic dual focus, oriented to both Greek and Roman culture in the Mediterranean area.

The view of artifacts and texts varied in eighteenth-century studies

Figure 2. The main gate to the Roman agora in Athens according to *Antiquities of Athens*
(1762) by James Stuart and Nicholas Revett (here taken from the German translation,
Stuart and Revett, 1829–1833, pls. 1 and 3). The two pictures show clearly how Stuart and
Revett presented the results of their measurements in different ways: by means of a
picture of the actual state of the ruins in the 1750s (left), and by reconstruction drawings
including data on dimensions and proportions, for further application by contemporary
architects (right).

of antiquity. For Johann Joachim Winckelmann (1717–1768), ancient
texts about classical art were a self-evident point of departure. In his
studies of ancient sculptures he sought "the perfect work of art" as a
model for contemporary artists. In his quest for this absolute aesthetic,
however, Winckelmann developed a formal stylistic analysis in 1764,
which enabled him to group ancient sculptures in five stylistic periods.
With the aid of ancient texts he placed ancient art in a rise-and-fall
model that was typical of his times: old style, high style, fine style,
imitating style, and decadence (Bianchi Bandinelli, 1978:34 ff.; Haskell, 1993:218 ff.; Schiering, 1969:1 ff.; Stoneman, 1987:110 ff.).

Architects in this period, on the other hand, placed a more obvious
emphasis on a primary knowledge of materials. Neoclassical architecture was based on detailed studies and measurements of Greek monu-

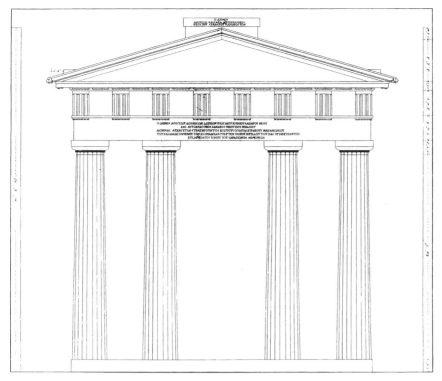

Figure 2. (*Continued*)

ments, especially temples, in Greece and the Greek colonial cities (Figure 2). One example is James Stuart's (1713–1788) and Nicholas Revett's (1720–1804) famous architectural studies, based on several years of travel in Greece, 1751–1754 (Schiering, 1969:55 ff.; Stoneman, 1987: 110 ff.). The importance of artifacts was also stressed by Count de Caylus (1692–1765), who introduced graphic pictures in publications of ancient artifacts. He viewed accurate depiction as a way to compare artifacts and their mode of manufacture. With a view of artifacts inspired by the natural sciences, de Caylus hinted that artifacts could sometimes tell us something different about the past from what we learn from texts. He thus touched on an idea that ran counter to the antiquarian premise that all artifacts must necessarily be associated with texts (Haskell, 1993:180 ff.; Schnapp, 1993:240 ff.).

With romanticism, the idea of evolution, and the emergence of European nation-states after the Napoleonic Wars, the interest in antiquity gradually changed character. Neoclassicism lost its dominating role and became merely one of many historicizing styles. At the same time, however, studies of antiquity were institutionalized by means of new public art museums and by educational reforms that put the classical ideal of education—above all, the classical languages—in the center. The clearest expression of the institutionalization of classical studies can be seen in Prussia, where school and university education were reformed in 1810, and where the Altes Museum in Berlin was opened in 1830. In a similar way, all the major European countries tried to portray themselves as the cultural heirs of antiquity—and especially of ancient Greece—by founding museums of "fine arts" and by giving a central place to classical studies in education. This increasingly clear definition of ancient Greece as the cultural cradle of Europe was made at the same time as modern Greece was created by the policy of the great European powers toward the Ottoman Empire (Bernal, 1987: 317 ff.; Schiering, 1969:67 ff., 94 ff.).

In contrast, the acquisition of ancient works of art by the large national museums was not institutionalized in the same way as museums and educational systems. A major part of today's European museum collections are instead the result of an unorganized hunt for works of art, largely on private initiative. A prime example is the frieze of the Parthenon, which Lord Elgin removed in 1801 and brought to London, but the treasure hunting can also be exemplified by the outright grave robbing outside the Etruscan city of Vulci, where 3,000 Greek vases were dug up in the years 1828–1829 alone (Cook, 1972:287 ff.; Stoneman, 1987:165 ff.).

The view of artifact and text in the first half of the nineteenth century was partly different from that of the eighteenth century. Romanticism placed the focus on language, and philology became the model science in the humanities (Bernal, 1987:224 ff.). Artifacts were in large measure subordinated to philological problems, but at the same time the textual criticism of the philologists meant that writing was perceived in a different, more critical way. Problems that received a great deal of attention were the identification of ancient sculptures, the iconographic interpretation of Greek vase paintings, and not least the ancient topography of Greece. The British officer William Martin Leake (1777–1860) solved most of the topographical problems on the Greek mainland in the 1810s, and German philologists charted the ancient geography of Asia Minor in the 1830s. The obvious point of departure for these topographical studies was ancient texts, especially Pausanias's description of Greece from the first century A.D. Yet the topographical reconstructions simultaneously required a thorough knowledge of ancient ruins and their position in the landscape (Stoneman, 1987:136 ff.).

The philological perspective is perhaps most evident in Eduard Gerhard (1795–1867), who sought to create a professional "philological archaeology" with the aid of the internationally imprinted Instituto di Corrispondenza Archaeologica, founded in Rome in 1828. Gerhard's view was that classical archaeology could not have a scientific basis until it used the methods of philology. Just as philologists and historians issued critical editions of documents, so classical archaeologists should publish the artifact collections of the great museums (Schnapp, 1993: 304 ff.). For Gerhard, who set the style with his large museum catalogues, the purpose of classical archaeology was primarily to provide background knowledge for the philological study of antiquity. The discipline thus became the study of the realia accompanying the historical study of language. The obvious starting point for Gerhard and his contemporaries was the museum collections. That was where archaeological work began, whereas little attention was devoted to the actual retrieval of the ancient objects (Schnapp, 1993:308 ff.).

The ultimate professionalization of classical archaeology came with a change in the view of the retrieval of evidence, when excavation became an important part of archaeological work (Figure 3). The French *Expédition scientifique de Morée*, conducted in the Peloponnese in 1829–1830, had already stressed the significance of systematic collecting (McDonald and Rapp, 1972:9 ff.), but the real breakthrough for archaeological excavations did not come until the 1860s and 1870s, when the industrialized great powers of Europe also began to institutionalize excavation. Special archaeological institutes were founded in Athens and Rome as bases for fieldwork. The retrieval of ancient artifacts was no longer based on short-term initiatives by individual fortune seekers but was, with few exceptions, given the form of state-organized and financed excavation projects that could last for decades (Schiering, 1969:44 ff.; 122 ff.). Modeled on the industry of the day (see Dyson, 1993), the excavations were organized in different stages and spheres of responsibility. The new significance ascribed to the actual uncovering of objects was evident in the way that all the major excavations were carried out with stratigraphic methods. In particular, Giuseppi Fiorelli (1823–1896) in Italy and Wilhelm Dörpfeld (1853–1940) in Greece developed and spread the principles of stratigraphy (Rumpf, 1953:92 ff.; Schiering, 1969:122 ff.; Stoneman, 1987:165 ff.).

Classical archaeology emerged as a separate discipline heavily influenced by earlier traditions in the study of antiquity. Alexander

Figure 3. Heinrich Schliemann's investigation of the grave circle at Mycenae in 1876 (section of a panoramic view, Schliemann, 1878:174–175). Although Schliemann financed his excavations from his private fortune, they represent the same kind of large-scale, long-term, systematic investigation as the contemporary state-financed archaeological projects in Greece.

Conze (1831–1914) maintained in a definition of the subject in 1869 that "Where the cross-section of classical philology and the longitudinal section of art history intersect, there, and precisely there, lies the field of classical archaeology" (cited from Niemeyer, 1968:27).

An overall question in the initial phase of classical archaeology was the problem of the origin of ancient art and architecture. Greece was seen as the obvious source of ancient art. The prevailing attitude, especially in Germany, was that the stimulus for classical Greek art came with Indo-European conquerors. In 1870 Conze formulated the idea of the "geometric" Greece, the style of which was associated with the presumed immigration of "primitive" Dorians from Northern Europe (Cook, 1972:287 ff.). Greek art was thus not just the origin of European art but also part of the common Indo-European heritage. The discovery of Mycenaean culture was an obvious problem for this line of thought. To rescue the Indo-European idea, Mycenaean culture was viewed either as the result of earlier invasions from the north or as "pre-Greek," a notion that was possible before the decipherment of the Mycenaean script, Linear B. Yet this "European" definition of Greece did not go unchallenged. Assyriologists and Scandinavian prehistoric and classical archaeologists especially drew attention to the strong oriental features and impulses in Greek art, although they often stressed the decisive Greek "renewal" of these impulses (Rahtje and Lund, 1991; Trigger, 1989:160).

In the manner in which ancient art history was pursued, classical archaeology was so strongly tied to ancient texts and philological textual criticism that Bianchi Bandinelli (1978:49 ff.) claims that the subject can be described as a continuation of "philological archaeology" until the First World War. The starting point for the major excavations was normally texts. Archaeologists chose above all to excavate sanctuaries and temples that were well known from written sources and that could be expected to contain important works of art (Rahtje and Lund, 1991). The German excavations at Olympia and the French campaign at Delphi are typical of this tradition, since the sites were "Panhellenic" sanctuaries that Pausanias described as being full of "masterpieces" of Greek art. During the first month's digging at Olympia, the German archaeologists did indeed find 42 sculptures of marble and bronze (Stoneman, 1987:263).

When profane urban settlement gradually attracted attention, written sources were likewise the point of departure. In the 1890s, German archaeologists began the first serious excavations of complete ancient Greek cities, initially Priene and Miletus on the coast of Asia Minor (Rumpf, 1953:92 ff.). Both were known from ancient sources for their early regular town plans, so the aim of the excavations was primarily to contribute to the art historians' discussion of the Greek art of town planning.

For Heinrich Schliemann (1822–1890) and later Arthur Evans (1851–1941) too, texts served as a basis for archaeological work, even if the texts were mythological (cf. Figure 3). Schliemann therefore searched for Troy with a copy of Homer in his hand, even though German textual critics had long since rejected the historical background of the *Iliad*. Schliemann's paradox was that his "incorrect" hypothesis led to excavations that uncovered a previously unknown preclassical literate culture (Daniel, 1981:98 ff.; Stoneman, 1987:265 ff.). Much later, with the decipherment of Linear B in the 1950s, this culture would also become a part of historical archaeology.

Yet even though the major excavations were based on ancient texts, the results were often more than just a simple confirmation of the written sources. Completely unknown sculptures were found, which showed that classical texts about art were written from special perspectives whereby certain works were mentioned while others were not (Bianchi Bandinelli, 1978:71 ff.). The great excavations also led to an explosion in the volume of material, particularly everyday objects, "whose 'classicism' could rightfully be doubted" (Niemeyer, 1968:27). These artifacts that were rarely or never mentioned in classical texts therefore had to be ordered chronologically and geographically on the basis of normal archaeological methods such as stratigraphy and typology. Many of the studied sites, moreover, proved to have a much longer history and a more complex topography than was known from ancient commentaries, such as Pausanias's description of Greece. In many cases, then, the great excavations led to the questioning of ancient texts (Bianchi Bandinelli, 1978:71). However, this was more a matter of questioning individual passages than of adopting an attitude of critical principle to the relation between material culture and writing.

Classical archaeology in the interwar years contained several contradictory tendencies. On the one hand, the period was one of "normal science," but on the other hand, it marked both an incipient departure from the view of antiquity as an exemplary ideal and an extreme exploitation of the classical idea. The normal scientific trend meant above all that given methods were refined. In the 1920s and 1930s more emphasis was placed on stratigraphy and find contexts, as is particularly clear in the Italian excavations at Pompeii and in the large-scale American excavations such as those at Pylos, Corinth, and the Agora in Athens (Morris, 1994; Skydsgaard, 1970). In the same way, considerable effort was expended on determining the chronology and origin of artifacts. Following the analogy of the art history of later periods, many objects were attributed to anonymous masters. John Beazly (1881–1970) in particular took this tradition to its extreme. He attributed about 30,000 Greek vases to over 600 anonymous masters, whom he mainly defined himself. These studies were thus a form of art history, without the names of the ancient artists, but peopled by masters with names like "the Berlin painter" and "the Florence painter" (Hoffmann, 1979).

At the same time, however, there were signs that the basic idea of antiquity as a given example was beginning to be questioned. With the modern breakthrough around 1900 in architecture, art, and literature, the need for historical examples was rejected. This perspective is clear in the works of Alois Riegl (1858–1905), who challenged the traditional rise-and-fall model deriving from Winckelmann (Bianchi Bandinelli, 1978:117 ff.). Riegl's ideas were carried forward in the interwar years by Ranuccio Bianchi Bandinelli (1900–1975), who dethroned the "eternal" Greek art by viewing it as a craft firmly anchored in its contemporary social and economic context (Bianchi Bandinelli, 1978:142 ff.; d'Agostino, 1991). Generally speaking, greater attention was paid to "nonexemplary" periods and phenomena, such as Greek prehistory, archaic art, and late antiquity.

A diametrically opposed trend was the extreme exploitation of classical ideas in Nazi Germany and fascist Italy. In these cases the exemplary nature of antiquity was emphasized by the direct association of the excavation, uncovering, and restoration of ancient monuments with political propaganda. Nazi Germany was particularly fond of the notion of the affinity of the "Indo-Germans" with Greece. This link was made especially clear at the Olympic Games in Berlin in 1936. On Hitler's direct orders, the German excavations at Olympia were resumed in 1935, with the aim of finding the then unknown stadium. The opening ceremony saw the introduction of a new custom, when the Olympic flame was borne by athletes running in relay from the German excavations at Olympia to the new Olympic stadium in Berlin (Jantzen, 1986). In Italy Mussolini's dreams of empire found expression in the demolition of parts of Rome with medieval and Renaissance buildings to expose fora from the imperial era and to build the magnificent Via dell'Impero. Propaganda equated Mussolini with Augustus, so Augustus's altar of peace, Ara Pacis, was likewise excavated and restored. At the same time, the exhibition city EUR created a new center for Rome, evoking clear associations with ancient Rome (d'Agostino, 1991; Frandsen, 1993). The close collaboration between many Italian classical archaeologists and the fascist regime gave classical archaeology such a political taint that very few archaeologists in Italy bothered with the subject in the years directly after the Second World War (Potter, 1987: 8 ff.; Rahtje and Lund, 1991).

With the reconstruction and modernization of Europe after the Second World War, the conditions for classical archaeology have become completely different from what they were when the discipline was created. In 1953 Andreas Rumpf could still approvingly cite Schopenhauer's declaration that "the Greeks are and remain the polestar for all our efforts, and the ancients will never be antiquated" (Rumpf, 1953: 135). Yet the radical economic and social changes in the postwar era have meant that the classical ideal of education espoused by Rumpf has gradually become an anachronism. The classical languages are no longer a natural part of a person's education, and references to antiquity play a very minor role in public life. In short, classicism has disappeared, and without classicism, classical archaeology has been marginalized. A great deal of the history of the subject since the Second World War reflects this dilemma: What is classical archaeology without classicism?

Broadly speaking, three answers to that question can be discerned (cf. Morris, 1994). The first main line consists in maintaining the original "great tradition" with little regard for the fact that the world has changed. Detailed stylistic descriptions of ancient works of art continue to be published, although these works no longer serve as examples. This conservative tradition is a history of ancient art that derives its inspiration from the great era when the subject was being built, but that is paradoxically isolated from recent debate in art history. And it is this tradition that has been the target of much of the criticism leveled against classical archaeology (cf. Dyson, 1993; Hoffmann, 1979; Humphreys, 1978; Morris, 1994; Renfrew, 1980; Snodgrass, 1985a, 1985b).

Another main trend consists in preserving but radically changing the art-historical perspective. The focus is no longer on descriptive, chronologically oriented studies but on such fundamental questions as the meaning and active role of the pictures. Representatives of this trend are Paul Zanker (1987, 1988), who has studied the expressions of Roman imperial power in images and architecture, and the so-called Paris School, which has analyzed the grammar of Greek vase paintings with inspiration from structuralism and semiotics (Bérard et al., 1989).

The third main line is the most radical breach with tradition in classical archaeology. Instead of pictorial art, the emphasis is on the economic and social history of antiquity. With impulses from history, anthropology, and prehistoric archaeology, traditional categories of finds, such as graves or ceramics, have been used to write an alternative "archaeological" history. The early history of Greece in particular has been studied this way, but work has also been devoted to Roman and Hellenistic Greece (see Morris, 1994). Moreover, with the decipherment of Linear B in the 1950s, a new historical archaeology was created in Greece that was less dependent on the older classicist ideals. Influences from prehistoric archaeology and anthropology have therefore been evident in studies of the Greek Bronze Age (Bintliff, 1977; Renfrew, 1972; Renfrew and Cherry, 1986).

The economic and social trend in classical archaeology is particularly clear in landscape archaeology. The interest in the landscape as a whole is mainly a postwar phenomenon (Figure 4). In Italy the first landscape surveys were carried out under the leadership of John Ward-Perkins, in southern Etruria in the 1950s (Potter, 1979). In Greece, landscape archaeology was introduced in earnest by the American Messenia expedition in the 1960s (McDonald and Rapp, 1972). In the 1970s and 1980s this kind of intensive survey had become an integral part of classical archaeology, having developed into a distinct form of fieldwork, with its own methods, problems, and potential (Barker and Lloyd, 1991; Keller and Rupp, 1983).

The breach with traditional topographical research in classical archaeology as a result of landscape archaeology is striking. Instead of identifying and studying individual places known from written sources, archaeologists study entire landscapes and all kinds of settlement, regardless of whether they are mentioned in ancient texts. In addition, landscape archaeology works with long temporal perspectives, from the Neolithic to the Middle Ages or the present day, with no special concern for any "exemplary" periods.

As a result of landscape archaeology, investigations of settlements such as Roman villas and studies of pottery, archaeologists have joined in the lively debate about the character of the ancient economy in a way that was previously unknown. There has long been a controversy between substantivists, who stress the different, "primitive" character of the economy, and formalists, who emphasize the "modern" features (Greene, 1986; Woolf, 1992).

The marginalization of classical archaeology has thus been handled in three different ways: by conserving the tradition, by renewing it, or by transforming it. It is interesting that the latter two ways have made the discipline less special, less "elect." Instead of retrieving an exemplary culture, classical archaeology has become one speciality concerned with one of many cultures in human history. It is precisely on the path away from the elect position that the relation between artifact and text has become the subject of theoretical discussion. It is only since archaeologists began to claim that they are writing alternative history, on the

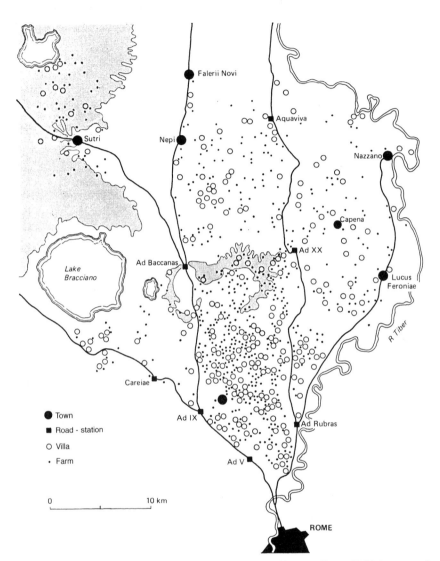

Figure 4. Settlement in southern Etruria around A.D. 100, according to British surveys in the 1950s and 1960s (Greene, 1986, fig. 104, which is based on Potter, 1979, fig. 35, by courtesy of Kevin Greene). The figure illustrates one of many results from the first systematic survey in the Mediterranean area.

basis of artifacts and pictures, that the relation between material culture and writing has become visible and been problematized.

In the debate of recent decades we can discern several different attitudes to the relation between artifact and text. For the ancient historian Moses Finley (1912–1986) the roles are obvious: "What is it that the classical historian ought to be asking of archaeological evidence today, and how successfully are the archaeologists, from their side, adjusting their own older aims and techniques to these new demands?" (Finley, 1971:170). For Finley the ancient texts are primary, so it is "self-evident that the potential contribution of archaeology to history is, in a rough way, inversely proportional to the quantity and quality of the available written sources" (1971:174 f.). He views numismatics and the history of technology, trade, and the Greek colonization of the Mediterranean, in particular, as areas in which archaeology can make significant contributions (Finley, 1971; see also Lloyd, 1986).

Anthony Snodgrass, who considered the issue on several occasions during the 1980s, uses less drastic words, but he still believes that the role of archaeology is greatest on the temporal and spatial periphery, that is, in early historical periods and in marginal, less well known areas. It is also striking that, like his pupils, he has mainly worked with archaic Greece. Yet Snodgrass also points out that archaeology can make independent contributions in central historical research fields too, such as political history. On the other hand, one should not expect any equivalence between artifacts and texts, since they were created in past activities with very different "levels" and "scales." Large agreements between written sources and material remains thus suggest circular arguments. Although Snodgrass asserts the intrinsic value of artifacts, he calls for a critical attitude among archaeologists to their own source material. He warns of "the positivist fallacy," the belief that what is immediately visible is also immediately important. He criticizes students of pottery, in particular, for their "grotesque" exaggeration of the commercial significance of ceramics (Snodgrass, 1983, 1985a, 1985b, 1987; cf. Vickers, 1990).

Philippe Bruneau (1974) likewise stresses the basic differences between artifact and text, but he sees the background more in their different relations to scholarly narrative, which is always in the form of text. Artifacts therefore have to be translated in a more fundamental way than texts. Stephen L. Dyson (1995) has recently taken up the translation problem (under the influence of postmodern literary criticism) when he emphasizes that classical archaeologists are authors who write texts after reading material culture, ancient written texts, and other present-day archaeological texts.

Comprehensive criticism of classical, and particularly Roman, archaeology has also been put forward by Eleanor Scott (1993), who calls for new research areas such as gender. She is critical of the often unspoken dependence on texts that she sees in Roman archaeology. She also points out that texts are material culture and, like artifacts, must be interpreted in relation to the social and symbolic contexts of the times when they were produced. A crucial factor for the character of Roman archaeology is that Roman texts cannot be regarded as representative of the empire as a whole. Archaeology could help to emphasize the "otherness" of the Roman Empire, that is, the parts of the empire that are unknown or alien to us.

A more two-way approach can be detected in studies where ancient texts are related to their physical context. These studies instead stress the link between the meaning of the text and its spatial—or archaeological—placing. This perspective has had an impact, above all, on studies of the Greek Bronze Age, since the Linear B inscriptions are found in archaeological contexts. John Bennet (1984) points out that texts have both material and nonmaterial aspects, since they consist of linguistic concepts reproduced on a physical surface. These two aspects of a text can be studied in broader contexts. It is thus important to distinguish between textual contexts, such as the content of written sources, and archaeological contexts, such as the circumstances in which texts are found. The link between a text and its spatial placing can in certain cases also give a new meaning to linguistic concepts. Several studies of the Mycenaean palace in Pylos have shown how Linear B texts analyzed in relation to their find context can lead to a deeper interpretation of the palace and its function (Bennet, 1985; Palaima and Shelmerdine, 1984; Palaima and Wright, 1985).

The relation between buildings, pictures, and texts has long been a subject of attention in the case of Pompeii as well, and new interest has been aroused in recent years (Laurence, 1994; Wallace-Hadrill, 1994; Zanker, 1988b). An example of an integrated analysis is Ray Laurence's reconstruction of neighborhoods (*vici*) in the city. He is able to detect this internal urban organization by comparing the position of street altars and public wells with painted inscriptions marking different electoral districts in the city (Laurence, 1994:38 ff.).

A two-way approach can also be detected in philological studies of classical epigraphy. Rosalind Thomas (1992) has relativized the written word by emphasizing oral culture in ancient Greece. With this perspective, the spatial placing of the inscriptions takes on crucial significance for the interpretation of the texts. In certain cases, such as the very brief inscriptions on boundary stones in Attica, it is almost impossible to interpret the text unless its spatial placing is known (Ober,

1995). The link between different forms of expression has also been studied by Jesper Svenbro (1989), who has carried out a close-up reading of an archaic female statue from Attica. He is able to show how the inscription, the woman's gestures, and her iconographic attributes allude to each other, like a game with words. Artifact, text, and image are thus seen as a whole that must be interpreted together.

With the questioning of the traditions of the subject, classical archaeologists have also begun to discuss the character of the discipline in earnest. Ian Morris (1994:45) makes a plea for "an integrated Greek historical archaeology," particularly devoted to studies of long time perspectives. Nick Kardulias (1994:39) similarly argues that archaeology in Greece should concern all periods, including the Turkish era, so that it can become "an anthropological historical archaeology."

The Christian Golden Age: Medieval Archaeology

Medieval archaeology can be seen as the shadow cast by classical archaeology. It is both its antithesis and its copy. Like classical archaeology, the foundations of medieval archaeology ultimately go back to the Italian Renaissance, but in a more negative way. It was then that the concept of the Middle Ages was created as a pejorative term for the "period of decline" separating antiquity from the Renaissance scholars. Yet it was not until the end of the seventeenth century and during the eighteenth century that the Middle Ages became an accepted designation for the period in Europe (Wienberg, 1993:180 ff.).

The period designated by the term Middle Ages is rather elusive and chronologically fluid. It was created as an antithesis to antiquity, but it has also come to be used in parts of Europe that have no ancient history. The general meaning of the concept comes closest to "Christian Europe," as opposed to the pagan Roman Empire and the pagan Germanic and Slavonic Europe. The chronological framework thus varies greatly from region to region. Moreover, the term is primarily applied to Catholic Western Europe, which came to an end with the Renaissance and the Reformation. For orthodox Christian Europe, the concept of the Middle Ages is less relevant. It is nevertheless used sporadically about the Balkans and Russia to denote the period between, on the one hand, antiquity and Slavonic paganism, and, on the other hand, the emergence of the Ottoman Empire and the Grand Duchy of Moscow (e.g., Fine, 1983, 1987). In the former Soviet Union, the Middle Ages were included in the chronologically broader concept of feudalism. The Middle Ages were therefore studied within the archaeological department of feudalism, established in 1934 (Trigger, 1989:228).

Studies of medieval monuments began even before the "Middle

Ages" had become an accepted period name. Seventeenth-century antiquarians in Northern Europe developed an interest in "medieval" churches, as one of several types of monuments from the past. In Sweden, churches were depicted alongside burial mounds and runic stones (see Figure 29), and in England John Aubrey (1626–1697) established the first typology of "medieval" architecture in 1670 (Andersson and Forsström, 1983; Schnapp, 1993:190 ff.).

The direct origin of medieval archaeology, however, must be sought in the romantic period of the early nineteenth century. That was when the Middle Ages were totally reappraised to serve as an example, just as classical antiquity had done. Europe thus acquired two exemplary epochs, antiquity and the Middle Ages; particularly "sublime" periods within these epochs were the fifth century B.C. and the thirteenth century A.D. As examples, however, the two periods functioned in different ways. The romantics placed the emphasis on language, the landscape, and the specific, which can thus be seen as the intellectual background to nationalism in Europe after the Napoleonic Wars. The quest for the Middle Ages thus became a quest for national identity, in a way that the pursuit of antiquity had not been (Kåring, 1992). A typical expression of the nationalistic features of medieval studies is that historians and students of monuments and historical styles stressed the greatness of their own countries. An illuminating example is Denmark, where Romanesque architecture from "The Great Age of the Valdemars" attracted considerable attention, whereas the Gothic styles of the German-dominated Late Middle Ages were not considered interesting (Wienberg, 1993:32 ff.).

Like classical studies, medieval studies were closely associated with aesthetics and philology. Medieval art and craft became obvious exhibits in the newly founded central museums of "fine art." For a museum man like Christian Jürgensen Thomsen (1788–1865), it was natural that his famous "Guide" (1836) should also include chapters about the Middle Ages and medieval artifacts. In Germany a special museum was even established for medieval works of art in 1852, Germanisches Nationalmuseum in Nuremberg (Fehring, 1991:1 ff.).

With the Middle Ages as an aesthetic model, historicizing styles, such as neo-Gothic and neo-Romanesque, were created, like the neoclassicism of the previous few generations. Some people thought that the neo-Gothic literally revived the Middle Ages, which was seen as a Christian golden age, characterized by Christian faith and morality (Kåring, 1992:37 ff.). To an even greater extent than neoclassicism, the medieval styles were based on stylistic purifications, since they were an expression of an age when the idea of evolution had its breakthrough

(Toulmin and Goodfield, 1965). These pure styles required detailed studies of the architecture and masonry of the surviving monuments. One example is the architect Thomas Rickman's (1776–1841) division of the English Gothic into Norman, Early English, Decorated, and Perpendicular styles in 1817. As Warwick Rodwell (1981:25) has pointed out, Rickman's classification was based on what we today would call archaeological principles, namely, typology and stratigraphy.

At the same time that new buildings were being erected in neo-Gothic and other historical styles, many of the medieval models were being restored. A wave of restoration swept over Europe, where some complete towns, most of the major cathedrals, and many smaller churches were mercilessly restored (Figure 5). In Britain alone, over 7,000 churches were restored in the period 1840–1873 (Kåring, 1992:51). Like the new buildings erected at the time, the restorations aimed for purity of style, which meant that the oldest or the predominant style of the monument was highlighted and that later additions and changes were removed; at the same time, the monuments were often stripped of their immediate surroundings. The chief proponents of stylistic uniformity believed that the aim of restoration should be to re-create the original architect's intentions, rather than what was actually built. With this idealistic attitude, the boundary between restoration, reconstruction, and new creation therefore became highly fluid (Kåring, 1992:187 ff.).

The principle of restoration may be compared directly with the contemporary philological principles for critical editions of texts. Just as the published texts sought to reconstruct lost originals on the basis of more or less corrupt series of copies, the restorations aimed to reconstruct an imagined original on the basis of more or less well preserved fragments of a building. Like the concentration on the big, famous monuments, which could often be associated with the "great" personalities in political history, the idealistic restoration principle can also be compared with the state-idealism of contemporary historiography. State-idealism was concerned with political history, shaped by "great" ideas and "great" men.

Several leading architects of the time worked with both restorations and new creations, and some of them also collected their experiences in historical surveys of medieval art and architecture. The practical application of the Middle Ages thus resulted in basic material knowledge of medieval monuments. Examples are Eugène Emmanuel Viollet-le-Duc (1814–1879), who published a 10-volume work on medieval French architecture, and Carl Georg Brunius (1792–1869), who wrote several works on Swedish medieval architecture. The bond be-

Figure 5. Cologne cathedral in 1887 (Dohme, 1887:218–219). The cathedral in Cologne represents the greatest "restoration" of a medieval monument in Europe in the nineteenth century. The Gothic cathedral was begun in 1248, but when construction stopped in 1560, only the chancel was completed, while the transepts, the nave, the aisles, and the towers were unfinished. On the basis of the medieval parts, the church was completed in pure High Gothic style 1842–1880. The huge building project was supported by all the German states, as an expression of "God, art, and the fatherland." When the cathedral was finally dedicated in the presence of Kaiser William I, the building symbolized the united Germany after the defeat of France in 1871 (Kåring, 1992:250 ff.).

tween these studies and medieval aesthetics and philology is particularly clear in Brunius, who was both a classical philologist and a practicing architect (cf. Grandien, 1974; Kåring, 1992:187 ff.).

As in classical archaeology, the retrieval of medieval artifacts was systematized in the second half of the nineteenth century. Fieldwork was mainly geared to collecting artifacts and compiling inventories of surviving buildings and ruins. In Germany, systematic publication of all historical—especially medieval—monuments began in the 1860s, and in Denmark, parish-by-parish surveys of both prehistoric and medieval monuments began in 1873 (Johannsen, 1992). It was also in this phase of systematized medieval studies that the first efforts to establish a professional medieval archaeology can be detected. This is clearest in the work of Hans Hildebrand (1842–1913), who claimed that archaeological studies of the Middle Ages had an intrinsic value (see Figure 33), partly because they could lead to the reinterpretation of medieval texts (Hildebrand, 1882).

The similarities between ancient and medieval studies are thus striking, but medieval archaeology was nevertheless not professionalized in the same way in the nineteenth century. I believe that the difference in development between classical archaeology and medieval archaeology has to do with accessibility in relation to the research perspectives. Ancient cities had disappeared, and ancient monuments were in ruins. Ancient art and architecture were therefore not directly accessible, so excavation was required for any detailed study. Medieval Europe, on the other hand, was preserved above ground to a much greater extent. Many cities still had a distinctly medieval character. The most important monuments were preserved and still functioning. Medieval art and craft were accessible in museums and collections. With the aesthetic perspective of medieval material culture, then, there was no need for professional, excavating medieval archaeology.

The first swing in the view of the Middle Ages and its monuments can be detected around 1900. With the breakthrough of modernism in art and architecture, the historical models in aesthetics were rejected, and the historicizing styles lost their significance. At the same time, the view of restoration changed with the recognition of the entire building history of the monuments, including the settings surrounding them (Kåring, 1992:304 ff.). Instead of being aesthetic examples, the monuments began to be viewed as historical documents, which could reflect a long, complex history. With this shift of perspective, many countries began the publication of historical monuments, describing their entire history. Several of the scholars who embraced the conserving principle of restoration, such as George Dehio (1850–1932) in Germany and Sigurd Curman (1879–1966) in Sweden, also took the initiative for the new studies of monuments (Kåring, 1992:304 ff.; Unnerbäck, 1992).

When the medieval monuments lost their direct modern-day function, as aesthetic models, there was instead an increased interest in the function and meaning the monuments had in the past. Since around 1900, therefore, an important tradition in architectural history has been focused on the study of the iconography of medieval architecture, that is to say, the meanings of the buildings. The meaning of church architecture in particular has been studied through comparisons of medieval texts about architecture, the medieval buildings themselves, and the pictures and inscriptions associated with the buildings (Kleinbauer, 1992:cx ff.). Since ecclesiastical architecture in particular has attracted attention, this tradition has had points of contact above all with the archaeology of medieval churches.

Systematic archaeological investigations of medieval remains also began around the turn of the century. This work was often a complement to preserved and known environments, in that the excavations concerned vanished or nonfunctioning sites, such as ruins of churches, monasteries, and castles, and towns that had disappeared or been moved. The excavations of ruins often sought to expose the monuments and hence make the past visible and accessible to the public (Andersson and Wienberg, 1993; Clarke, 1984). In those cases where archaeology was conducted in existing towns, such as Lund in present-day Sweden, it is typical that the excavations concerned places with a past age of greatness.

In the interwar years, medieval archaeology underwent a gradual growth, especially as a result of excavations conducted by people with a background in history and art history. The work was intended primarily as a complement to the study of medieval texts. It was considered important to trace the oldest history and topography of individual towns (Redin, 1982), or to shed light on the building history of individual monuments (Klackenberg, 1992:26 ff.), but less familiar aspects of the Middle Ages were also studied, such as early urban crafts (Blomqvist, 1941) and the agrarian economy (Jansen, 1984).

Although medieval archaeology existed as a practical activity in the 1920s and 1930s, it was not until after the Second World War that the subject was professionalized and became an academic discipline. In the 1950s and 1960s archaeological excavations really got under way in surviving medieval settings. That was when excavating archaeology and the architectural documentation of masonry were integrated into a complete stratigraphic analysis of buildings (see Figure 28) (Andersson and Wienberg, 1993; Rodwell, 1981).

Although the professionalization of medieval archaeology took place long after the Middle Ages became a subject of keen interest, the subject shares certain nationalistic features with the original romantic interest in the Middle Ages. The focus has often been on monuments and sites that can be associated with the medieval states and their emergence. An illustrative example is the establishment of Polish medieval archaeology. In Poland, which was newly created in both political and geographic terms after the Second World War, great archaeological attention was devoted in the 1950s and 1960s to the emergence of the Polish state. This work was expressly carried on as part of the official millennium of Poland, commemorated in 1960–1966 (Leciejewicz, 1980).

The reconstruction and rebuilding of European cities after the Second World War is often pointed out as a decisive factor for the growth of the subject (e.g., Fehring, 1991:1 ff.). The large-scale archaeological excavations since the 1950s in many medieval European cities have undoubtedly affected medieval archaeology and its character. More fundamental factors should be sought, however, in changed perspectives on both archaeology and history. Archaeology was no longer viewed as an excavating branch of art history, but rather as an extension of history, whereas history was increasingly concerned with social and economic history. Medieval archaeology has therefore been heavily influenced by history (Austin, 1990), and very few medieval archaeologists work with the traditional source material of art history, such as church murals, wooden sculptures, manuscript illustrations, and artifacts of gold, silver, enamel, and ivory. In other words, medieval archaeology studies a much more limited part of material culture than does, say, classical archaeology. One of the few medieval archaeologists who has had a broader perspective and preserved a link with the study of the iconography of architecture is Erik Cinthio (1957, 1968), who has devoted several works to medieval architecture and mural painting, where he has stressed the importance of the meanings conveyed by material culture, thus anticipating in part today's symbolic currents in archaeology.

The otherwise clear link with history means that research in medieval archaeology is traditionally text-bound and thematized according to the self-understanding of the Middle Ages—the doctrine of the four estates of society—in studies concerning the countryside, the towns, the churches, and the castles (Andrén, 1994). In addition, there are special studies of artifacts, such as pottery, but these are not at all as fundamental for the character of the subject as the division of materials in classical archaeology. The thematization is highly obvious in the internal working groups found in English medieval archaeology: the Deserted Medieval Village Research Group (1952), the Urban Research Committee (1970), the Moated Sites Research Group (1972), the Churches Committee (1972), and the Medieval Pottery Research Group (1975) (Hinton, 1983:83 ff.). A comparable internal specialization, with separate working groups and conferences, can also be detected in French, German, and Scandinavian medieval archaeology. It is also found in handbooks of medieval archaeology from a number of countries (e.g., Barry, 1987; Clarke, 1984; Fehring, 1991; Liebgott, 1989).

The interpretative perspective in medieval archaeology has not been very explicit. On the basis of a general idea of complementarity, archaeology has been seen primarily as a method for supplementing contemporary written sources. This perspective has meant that the archaeological interest in the Middle Ages has very different chronological centers of gravity in Europe. In Western and Southern Europe the study has mostly concerned the Early Middle Ages, since there are ample written sources from later periods (see Francovich, 1993). In Northern and Eastern Europe, on the other hand, where there are far fewer texts, the whole period has been studied with more equal intensity.

The idea of complementarity has also meant that many medieval archaeological investigations have had the character of detailed studies in relation to a given historical synthesis. Above all, aspects of the Middle Ages that are less well known from texts have been studied. An important area, as in classical archaeology, has been settlement, primarily in towns, but also rural settlement (see Figure 32). Although churches and castles have been studied as individual monuments, they have often been incorporated in the perspective of settlement archaeology. In the same way, for example, medieval iron production and medieval everyday life have been studied archaeologically, to compensate for the dearth of written sources dealing with these areas (Andersson and Wienberg, 1993; Felgenhauer-Schmiedt, 1993).

In the last fifteen years, however, there has been a renewal in the subject, in that the interpretative imperative of the written sources has been questioned. The renewal is particularly noticeable in Scandinavia and Britain, partly due to impulses from anthropologically inspired history, such as the *Annales* school and "the new cultural history," and partly due to the active integration of the debate in prehistoric archaeology and anthropology into medieval archaeology. The changed character of the discipline has been expressed in two partly different ways. One reaction has been to write more independent archaeological syntheses about major medieval problems (Figure 6), such as farming (Astill and Grant, 1988), the villages (Chapelot and Fossier, 1985), the towns (Andersson, 1990; Andrén, 1985), trade (Hodges, 1982a), craft (Christophersen, 1980), iron production (Francovich, 1989), coin circulation (Klackenberg, 1992), the churches (Morris, 1989; Wienberg, 1993), and mortuary practices (Redin, 1977). In this case, the renewal has been stressed by means of an emphasis on the role of

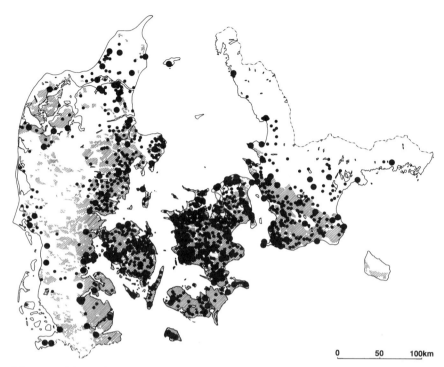

Figure 6. The gothicization of parish churches in late medieval Denmark, 1300–1550 (Wienberg, 1993, fig. 70, by courtesy of Jes Wienberg). Hatched areas show the occurrence of fertile clay till, while the dots show parish churches to which Gothic vaults were added, with the size indicating the number of vaults in each church. The map, which is based on an analysis of the 2,692 medieval parish churches in the area, shows how the Gothic rebuilding projects are mainly concentrated in the most important agricultural areas, which were dominated by aristocratic landowners. The study exemplifies how analyses of a kind of source material that attracted attention from an early stage, such as ecclesiastical architecture, can be used to study new questions, such as the relation between ideology and economy.

archaeology in connection with important problems that have long been debated by historians.

In another, more radical approach, the very idea of archaeology as a complement to history has been questioned. David Austin (1990) argues that it is not sufficient to provide archaeological perspectives on historical problems. For him it is precisely the basic link with history and historical issues that has hampered the development of method in medieval archaeology and the potential of studies of material culture. He calls instead for new questions cutting across the traditional thematization of the subject. An example of a new type of issue is the study of gender, which for Roberta Gilchrist (1994:8 ff.) means that artifact and text become more equivalent as source material. This attitude takes its inspiration from contextual archaeology with its emphasis on the meaning and active role of artifacts. Yet this inspiration has especially functioned as a way to link up once again with the similar but much older debate about the iconography of architecture (see Gilchrist, 1994:17).

As a stage in the gradual renewal of medieval archaeology in the last 15 years, both the character of the subject and the relation between artifact and text have been discussed intensively. However, because of the national fragmentation of the subject, this debate has been divided into different language areas and has thus been conducted more or less independently in, for example, Britain, France, Italy, Germany, Poland, and Scandinavia. The viewpoints and perspectives, however, are strikingly parallel. Archaeology has been perceived in all these countries as particularly suitable for the study of areas that are rarely or never mentioned in written sources, such as technology, economy, social conditions, and everyday life (e.g., Cinthio, 1984; Dymond, 1974; Jankuhn, 1973). Earlier than in other branches of archaeology, medieval archaeologists discovered the French *Annales* historians, since many of them are medieval specialists. Several scholars have pointed approvingly to Fernand Braudel's "long waves" as suitable objects of archaeological study (Biddick, 1984; Hodges, 1989; Klackenberg, 1992; Österberg, 1978). Yet there has also been criticism of the concentration of archaeology on economic and social questions. Wolfgang Seidenspinner (1989) claims that the archaeological emphasis on these questions is often based on an underestimation of the evidential value of the written sources. He argues instead that material culture is a special dimension in life and that archaeology therefore can just as well study mental and political issues.

The question of the role of texts in archaeological work has been perceived in very different ways. An earlier tradition stressed the given historical background knowledge. For Michel de Bouard (1975) the starting point for archaeology is always questions formulated in history. One reaction to this stance has been to reject written sources in general. This perspective has been particularly clear in the attempts to introduce a "new medieval archaeology" (Hodges, 1982b; Rahtz, 1981, 1983). For Philip Rahtz (1983:13), medieval archaeology is "a discipline with its own theoretical basis, which may arrive at conclusions of historical interest, independently of the evidence of written sources." Even among medieval archaeologists who were early to reject new archaeology, the intrinsic value of archaeology as opposed to texts has also been stressed (e.g., Christophersen, 1979). The most consistently conducted study of this pure archaeological tradition is Lars Redin's (1977) study of a cemetery in Skanör. Although the town was one of the most important marketplaces in Northern Europe and was therefore well known from written sources, Redin chose to make a purely archaeological interpretation of the cemetery, with no regard for contemporary texts.

Directly linked to the rest of the discussion is the issue of the relation between material culture and written sources. Roughly speaking, the two forms of expression are viewed as either basically the same or fundamentally different. Axel Christophersen (1979:6) maintains that artifacts and texts are in principle similar, even though they belong to "different segments—and different levels—of the objective past reality." In a similar way, Jes Wienberg (1988) claims that there is no decisive difference between artifact and text, since both are at once "remains" and "signs" from the past (see also Christophersen, 1992). Stephen Driscoll (1988) and John Moreland (1991, 1992) stress instead that all texts are in a fundamental sense artifacts and that both forms of expression should be interpreted together. In this way, writing does not have primacy of interpretation over material culture.

In contrast to these stances, Dymond (1974:20) stresses the difference between conscious texts and less conscious artifacts: "Written and spoken statements can express a range of intellectual and spiritual considerations, which material evidence, even works of art, could never even approach." An assumption of a basic difference between artifact and text is also a precondition for Reinhard Wenskus (1979), when he pleads that we should deliberately look for contradictions between material culture and written sources. These contradictions are important for him, since there is no definitive truth in studies of the past.

I have previously tried to distinguish both similarities and differences between artifacts and texts. On the one hand I have argued that writing in general can be seen as more conscious or manifest than material culture (Andrén, 1985). On the other hand, I have also pointed out that certain artifacts are more "textlike" than others, such as coins and formalized architecture. "Material culture" thus seems to be an umbrella term for very different things (Andrén, 1988).

Finally, the character of the subject itself has been discussed in detail in recent years. Some scholars stress the multidisciplinary role of the subject (e.g., Cinthio, 1963; Dymond, 1974) and the period itself— the Middle Ages—as particularly significant (Cinthio, 1984), whereas others would like to burst the chronological boundaries in various ways, to make the subject into a general historical archaeology (Andrén, 1988; Wienberg, 1988). Some would instead see medieval studies as part of general archaeological work (Bertelsen, 1992; Christophersen, 1992), and others view medieval archaeology as one of several subjects studying complex societies (Mogren, 1990; Tabaczyński, 1987).

Historical Archaeologies in the Middle East and Asia | 3

THE MIDDLE EAST

As in Europe, there are a number of different archaeologies studying the historical periods of the Middle East. We have Egyptology, Assyriology, biblical archaeology, classical archaeology, Arabian archaeology, and Islamic archaeology. Unlike the specialities in Europe, however, the historical archaeologies of the Middle East are not of indigenous origin. They were instead created in a Europe that started to dominate the Middle East politically and economically when the Ottoman Empire began to disintegrate. By amassing knowledge about the "magnificent" past of the area, the European great powers saw themselves as more worthy heirs of the old civilizations than the modern-day inhabitants. A knowledge of the region's past could thus justify European claims to power here (Said, 1978). The historical archaeologies concerning the Middle East are still in large measure dominated by European and American archaeologists.

The specialities were created out of an ambiguous European interest in the Middle East. On the one hand, the area was seen as the ultimate source of European civilization, not least as regards art and the biblical tradition. On the other hand, the area represented an unchanging "oriental" despotism and decadence, which many Europeans repudiated, partly as a projection of the opposition between Christianity and Islam. This construction of the unalterably different Orient was the antithesis of the contemporary idea of the ever-changing, dynamic Europe (Said, 1978).

Unlike the historical archaeologies in Europe, which mostly concentrated on material culture, the specialities about the Middle East were more often studies of geographic areas in which archaeology was integrated with linguistic studies. It is impossible, for example, to become an Egyptologist without a mastery of hieroglyphics, whereas many classical and medieval archaeologists have little knowledge of Latin or Greek. The link between language and archaeology in the Middle East is due to the fact that the emergence of the historical archaeologies was closely connected to the decipherment of the ancient written languages of the area.

All the historical archaeologies in the Middle East contain interesting studies and relevant discussions (see, e.g., Carter and Stolper, 1984; Grabar, 1971; Hole, 1987; Potts, 1990; Redman, 1986). In this context, however, I shall consider only the oldest and most distinctive traditions in the field, namely, Egyptology, Mesopotamian archaeology, and biblical archaeology.

The Origin or Cul-de-Sac of History: Egyptology

Interest in pharaonic Egypt goes back a long way. Greek writers regarded Egypt as the origin of culture. In the Roman period, Egyptian obelisks were removed to Rome and Constantinople, and the unreadable hieroglyphs were viewed as a source of ancient wisdom both by Romans and by the early church fathers (Bernal, 1987:121 ff., 161 ff.; France, 1991:1 ff.).

In modern Europe, however, there has long been a cleavage in the attitude to ancient Egypt. During the Enlightenment, Egypt was admired for its political system, but when Greece was defined in the first half of the nineteenth century as the origin of Europe, many people dismissed Egypt as a "cul-de-sac" in the history of civilization. It was seen as a "sterile" culture (Bernal, 1987:224 ff.). At the same time, pharaonic Egypt retained something of its attraction, as one of the countries in the Bible. Some people even thought that Egypt was the zenith of human development, after which everything degenerated. This idea is most clearly seen in G. Eliot Smith's and W. J. Perry's hyperdiffusionism from the 1920s, which saw all essential innovations in human history as having spread from Egypt (France, 1991:16; Reid, 1985; Trigger, 1989:152 ff.).

Despite the old interest in Egypt, it was not until Napoleon's campaign of 1798–1801 that pharaonic Egypt really became known in Europe. The campaign was at once an attempted conquest and a scientific expedition, modeled on Alexander the Great's conquests. According to Edward W. Said, the French campaign marked the beginnings of "orientalism," as a way for great European powers to rule the area by means of knowledge (Said, 1978:79 ff.). During the French occupation, monuments and inscriptions were registered, and tens of thousands of artifacts were collected and sent to Paris for future publication (Figure 7). The most important single find was the Rosetta Stone, discovered in 1799, which was the basis for the decipherment of hieroglyphs (France, 1991:7 ff.). The increased knowledge, typically enough for the times,

Figure 7. French scientists measure the sphinx at Giza (Denon, 1802, p. 20:1). According to Said (1978), the attempted conquest of Egypt by the French in 1798–1801 marked the start of "orientalism" as a way for the great powers of Europe to dominate the world through knowledge. In other words, the figure may be said to illustrate orientalism in practice.

quickly found a practical application in the Empire style, which spread from France over Europe in the first decades of the nineteenth century (Curl, 1982).

A decisive turning point in the study of ancient Egypt was the final decipherment of hieroglyphs in the 1820s and 1830s, mainly carried out by Jean François Champollion (1790–1832). Although parts of the history of pharaonic Egypt were known earlier, the decipherment opened up a largely unknown 3,000-year civilization, which became accessible through its own texts. After just a few decades, the chronology of the political history of Egypt was charted in broad outline (Gordon, 1968:19 ff.).

The French and later the British presence in Egypt aroused European interest in Egyptian works of art. From around 1815, many of the rich graves were plundered by diplomats and emissaries from Britain and France, often in more or less open competition. Art treasures were sold or handed over to the newly founded central museums in Europe. Despite the huge public interest in mummies and grave finds, however, the Egyptian artifacts were often given a less prominent position in the museums than classical art, because pharaonic Egypt was regarded as inferior to ancient Greece and Rome (France, 1991:27 ff.).

A few Egyptians expressed criticism of the British and French plundering as early as the 1830s, and the European attitude gradually changed in subsequent decades, although control of Egyptian artifacts remained in European hands (Reid, 1985). A Prussian expedition in 1842–1845 carried out the first systematic survey of monuments in Egypt, and French Egyptologists created an antiquarian organization in 1858 and an archaeological museum in Cairo in 1863. For the initiator, Auguste Mariette (1821–1881), the idea was that the museum would preserve as many as possible of the artifacts and works of art in Egypt, instead of letting them disappear to the national museums in Europe. The plundering and treasure hunting continued, but the museum did exert a certain degree of control over the flow of objects (France, 1991: 125 ff.). At the same time, however, Mariette, by virtue of his power over the museum, actively prevented some Egyptians from working with Egyptology (Reid, 1985).

The final professionalization of archaeology in Egypt did not take place until the 1880s and 1890s, particularly with the excavations of William Flinders Petrie (1853–1942). Petrie, who was opposed to plundering and treasure hunting, introduced stratigraphy, excavated secular buildings as well, and established pottery chronologies independent of the texts. He collected all finds, regardless of their "artistic" value, which led the British Museum to complain that Petrie brought home "a vast quantity of pottery and small objects which from our point of view are worthless" (cited from France, 1991:191).

A decisive factor for the orientation of professional archaeology in Egypt was that political and religious life in pharaonic Egypt had been clarified long before, with the decipherment of the hieroglyphs. For Flinders Petrie and his contemporaries, the purpose of archaeology was to study the origin of the civilization in predynastic Egypt, and to integrate the finds from the historical period in a given cultural framework (Kemp, 1984).

With the spectacular find of Tutankhamen's grave in 1922 and the sensational publication of Nefertiti's portrait in 1923, the position of archaeology in Egypt changed. The grave and the portrait became crucial in a nationalist struggle for power over the country's past. The first indigenous education in Egyptology began in 1923, and with partial self-government for Egypt in 1924, domestic control over excavation was also tightened. A ban on the export of ancient artifacts led to a great reduction in the number of foreign archaeological expeditions from the 1930s onward. It was not until Nasser's revolution in 1952 that Egypt acquired full administrative control over archaeology in the country, and today there are many indigenous archaeologists (Reid, 1985).

Although there has been an indigenous Egyptology since the 1920s, the view of the pharaonic period is contradictory in modern Egypt. For many Muslims, pharaonic Egypt represents a period of polytheistic idolatry that ought to be condemned. Interest in ancient Egypt therefore has its main base in the Western-influenced middle class, who have refined a sort of "pharaonic nationalism" (Reid, 1985; Silberman, 1989: 153 ff.). The country's Coptic minority in particular has taken an interest in pharaonic Egypt, since the last remains of the ancient Egyptian language are represented by the liturgical language of the Coptic church. Many Copts therefore feel that they are the lawful heirs of the Pharaohs, and many indigenous Egyptologists have in fact come from the Coptic minority (Reid, 1985).

After a long period with few foreign excavations, foreign archaeological presence in Egypt increased once again with the great Aswan project in 1960–1965, when 55 countries took part in the excavation of hundreds of sites and moved about 20 temples (Bratton, 1967:257 ff.). Since then, several large-scale foreign excavations have been carried out, but unlike older expeditions with their focus on graves and temples, the studies have mainly concerned profane buildings (Bierbrier, 1982; Wenke, 1989). It has not been possible, however, to introduce landscape archaeology on any scale because of the Nile's alluvial deposits and the intensive modern exploitation of the river valley. However, archaeology has been brought into the crucial discussion of the urban character of the pharaonic state. Until the 1970s, the traditional picture of Egypt was that the country was a civilization without cities. Yet archaeology has shown that Egypt was dominated by a small number of very large cities (Wenke, 1989). British excavations at El-Amarna (Figure 8) have shown that it was a big city, with buildings of sun-dried clay standing between the large, well-known stone temples. Archaeology has thus served as a direct challenge to traditional knowledge of Egypt (Kemp, 1977, 1984, 1989:84 ff., 261 ff.).

Despite this challenge, however, archaeology is still in large measure dependent on texts, because of the characteristic professional division of Egyptology into two parts. Ever since the professionalization of Egyptology, Egyptologists with a linguistic orientation have occupied academic posts and thus dominated the subject intellectually. Their focus on texts, pictures, and architecture has led to a concentration on philology, art history, and to a certain extent political history. On the other hand, Egyptologists with an archaeological orientation have occupied the museum posts so that they can handle the museums' artifact collections (O'Connor, 1990; Trigger, 1993:1 ff.). As a result of the obligatory language studies and the linguistic dominance, archaeologists have thus, in a double sense, worked "in the shadow of texts" (Kemp, 1984).

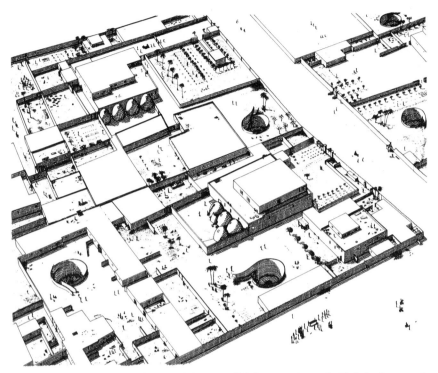

Figure 8. Reconstruction of parts of the city of El-Amarna, latter half of the fourteenth century B.C. (Kemp, 1989, fig. 98, by courtesy of Routledge). Archaeological excavations in El-Amarna have revealed extensive urban settlement between and around the great temples. In this case archaeology has functioned as an outright challenge to older, more text-based interpretations in which pharaonic Egypt has been viewed as a civilization without cities.

The linguistic preponderance, with the focus on the unique hieroglyphs, has also meant that the unique sides of pharaonic Egypt have been emphasized. The concentration on this uniqueness has led to scholarly isolation, which Bruce G. Trigger (1993:16) has tried to break by comparing Egypt with six other early civilizations. He argues that the unique and the general in a society cannot be determined *a priori*; they can only be identified after empirical comparisons.

Despite the philological dominance, questions about the relation between artifact and text have been touched upon in Egyptology. John Baines (1988) maintains that archaeological interpretations should also include writing, since it was so significant. Especially important was the large-scale and unique use of writing on monuments. To include writing in archaeological interpretations, however, the various uses of texts must also be clarified. Through studies of the first dynasties, Baines is able to show how writing was a direct expression of the culture and ideology of the elite, and how the use of texts varied with the concentration and division of political power. He believes that this varying use of writing should be compared with the material remains, so that "archaeology and writing complement each other's silences" (Baines, 1988:209). Another important aspect of the function of text was the deliberate destruction of inscriptions. Particularly during and after the attempts at religious reform by Akhenaton (ca. 1353–1335 B.C.), names were erased and entire inscriptions buried, in deliberate endeavours to rewrite history (Kemp, 1989:261 ff.).

Barry J. Kemp (1977, 1984, 1989) has also touched on the relation between artifact and text on several occasions. He emphasizes the close connection between writing, monument, and picture, especially in view of the emblematic form of the writing, which made it appropriate for wordplay and associations between text and text-bearing objects. Kemp claims that the writing together with the pictures and monuments has conveyed an idealized version of Egyptian society, which has above all been portrayed as unchanging: "Indeed the essence of the act of writing (and of drawing) is to reduce a complex and often chaotic reality to a comprehensible order" (Kemp, 1989:130). Against this ideal image, Kemp emphasizes profane settlement and the more everyday, nonformalized objects, which reveal many fundamental changes in pharaonic Egypt. For Kemp (1989:27 ff., 53 ff., 84 ff., 137 ff.), then, there is an important boundary, not between artifact and text but rather between formalized artifacts (including texts) and nonformalized objects. At the same time, he points out that there is a general difference between artifact and text that is especially cognitive. Written sources are verbal, like our modern intellectual culture, whereas artifacts, art, and architecture are more visual. This requires an "intellectual transformation" of archaeological data into text, but archaeology is poorly equipped for this transformation. Kemp therefore sees archaeology as "a flawed discipline," although he constantly stresses its significance (Kemp, 1984:27).

Paradise Lost: Mesopotamian Archaeology

Through the Bible and ancient historians such as Herodotus, part of the early history of Mesopotamia was known in Europe, but it was not until the seventeenth and eighteenth centuries that European travelers began to take an interest in the remains of the vanished empires. The undeciphered cuneiform script exerted a special fascination; the first accurate drawings of cuneiform inscriptions were made in 1765 by Carsten Niebuhr (1733–1815) (Lloyd, 1980:7 ff.).

As in Egypt, interest in Mesopotamia was concerned with the history of civilization. Scholars were interested in seeking the origin of European culture, not least the ultimate source of Christianity and Judaism. As in the case of Egypt, the attitude to Mesopotamia was ambivalent. It represented the lost paradise of the Bible, and some saw it as the cradle of all civilization, the principle being "ex oriente lux." One example is Hugo Winkler, whose pan-Babylonism claimed that all the myths in the world were more or less direct reflections of a system of narratives created in Babylonia around 3000 B.C. (Larsen, 1987). Others toned down the role of Mesopotamia in the history of civilization, instead emphasizing the unique and innovative significance of Greece (Bernal, 1987:237 ff.).

Firsthand knowledge of Mesopotamia did not reach Europe until the 1830s and 1840s. That was when French and British diplomats tried to gain control over Mesopotamia, which was part of the disintegrating Ottoman Empire. At the same time, these diplomats excavated—or pillaged—Assyrian and later Babylonian cities. The principle, as Austen Henry Layard (1817–1894) put it, was "to obtain the largest possible number of well-preserved objects of art at the least possible outlay of time and money" (cited from Lloyd, 1980:108). The finds, which included tens of thousands of cuneiform tablets, were brought to the central museums in London and Paris, where they attracted enormous attention (Figure 9). For the first time, remains from the biblical countries could be seen in Europe (Larsen, 1994; Lloyd, 1980:97 ff.).

Simultaneously with the first treasure-hunting excavations, the cuneiform script was deciphered. This writing system was used for several different languages in the Middle East, and the most important of these were deciphered between the late 1830s and 1860. The decipherment had begun around 1800, but the real breakthrough came with Henry Rawlinson's (1810–1895) studies of cuneiform texts in the 1830s and 1840s. In 1837 he was able to present the first translation of two sentences written in Persian (Larsen, 1994:317 ff.; Lloyd, 1980:73 ff.).

As with Egyptian hieroglyphs, the decipherment of Mesopotamian cuneiform meant that a whole new world was opened up, with 3,000 years of political history. However, the cuneiform texts shook the biblical tradition in a way that the hieroglyphs did not. The texts from Mesopotamia became important arguments in the heated religious debate that broke out after the publication of Darwin's *On the Origin of*

Figure 9. An English couple inspecting an Assyrian sphinx at the British Museum in 1850 (Larsen, 1994:241, after *The Illustrated London News*, by courtesy of Gyldendal). The finds from the pillaging excavations of the British and the French in the cities of Mesopotamia attracted immense attention when they were first exhibited in London and Paris in the late 1840s. Never before had Western Europeans been able to see works of art from the biblical countries.

Species in 1859. In several cases the texts revealed Assyrian and Babylonian parallels to narratives in the Old Testament; a Babylonian version of the Great Flood, discovered in 1872, attracted particular attention (Lloyd, 1980:144 ff.). The Bible could no longer be seen as a unique text; it was now just one of many expressions of a greater narrative tradition in the Middle East. In Germany especially, this relativization took on anti-Semitic features. For the Assyriologist Friedrich Delitzsch (1850–1922), author of *Babel und Bibel*, the Mesopotamian origin of biblical tradition meant that Christianity could be freed from its Jewish heritage and thus become the first "true universal religion" (Larsen, 1987).

The tremendous importance of the texts in Europe at that time meant that the incentive for continued excavation in Mesopotamia was the hope of finding new texts. As before, archaeology was characterized by more or less organized predatory digs, especially in Assyrian cities.

In the 1880s alone, between 35,000 and 40,000 cuneiform texts were spread over the world through the hands of antique dealers in Baghdad (Lloyd, 1980:162).

The final professionalization of Mesopotamian archaeology came around 1900, when large state-run excavation projects were started, first by Germany and later by the United States. The most famous dig was conducted in 1899–1914 in Babylon (see Figure 38), under the direction of Robert Koldewey (1855–1925). He was schooled in German classical archaeology, from where he introduced stratigraphic excavation methods, which meant that walls of sun-dried clay were for the first time discerned in Mesopotamia. For Koldewey the whole city was interesting, even though it did not yield any large amounts of text or splendid works of art. The monumental architecture nevertheless attracted attention, and a whole city gate—the Ishtar Gate—was brought to Berlin, where it was reconstructed after the First World War (Lloyd, 1980: 173 ff.).

In the interwar years, the tradition of studying the historically known cities continued, but certain changes can also be observed. The archaeological interest was extended from the biblically known Assyrian and Babylonian cities to other areas and earlier periods. The Sumerians, who had previously been known only indirectly, were now studied by British excavations in Ur. The study of truly prehistoric remains also began, with deep trenches being dug through tells (Lloyd, 1980:194 ff.). As in classical archaeology, archaeologists, particularly those working on American excavations, began to pay more attention to the find context (Ellis, 1983). Gradually, however, archaeological work was subject to tighter regulation in Iraq, which was then a British protectorate. A museum was established in Baghdad in 1926, and a ban on the direct export of ancient artifacts was imposed in 1932. As in Egypt, foreign archaeological activity in Iraq declined as a result of the tougher restrictions, and many archaeologists turned to Syria, from where finds could still be taken out of the country fairly easily (Lloyd, 1980:194 ff.).

The postwar period is marked by several major changes in Mesopotamian archaeology. Indigenous archaeology was established in earnest in Iraq in the 1940s. For this Iraqi archaeology, the relation to the history of European civilization is no longer of central importance; now there is more concern for issues of national identity and past greatness, roughly as in the case of the historical archaeologies in Europe. Another expression of the nationalism of Iraqi archaeology is the interest in later phases, especially the Islamic period (Lloyd, 1980:194 ff.). In the last few years, earlier periods such as the Babylonian era have also been actively used in political propaganda (see Chippendale, 1991).

From the 1950s onwards, large-scale foreign projects were resumed in Iraq, first by the Americans, later by the Russians. The projects have focused on the previously unknown countryside. As a result of major landscape surveys, villages and irrigation channels between the cities have been charted over long periods of time. Compared to the total focus of earlier traditions on "historical" cities, the changes of perspective brought by landscape archaeology are striking. Of the hundreds of places that have been localized by one American survey, only about 10 sites outside the cities have been identified through texts (Adams, 1981; Adams and Nissen, 1972).

Landscape archaeology has also served as a basis for more large-scale comparisons of the history of civilization. By juxtaposing Mesopotamia with Central America, Robert McC. Adams (1966) has sought for regularities that are historically independent of each other in the emergence of early towns. In his study, as in later similar comparisons (e.g., Trigger, 1993; Wheatley, 1971), it is important to have both ample written sources and comprehensive archaeological material. The condition also applies to the discussion of early world systems, of which Mesopotamia is sometimes reckoned as the oldest (cf. Kohl, 1989; Lamberg-Karlovsky, 1989).

On a more small-scale perspective too, certain changes can be detected in Mesopotamian archaeology. The archaeological context of cuneiform texts was considered as early as the 1930s, but it is only in the last 15 years that this context has also been actively used for integrated interpretations of settlement (Ellis, 1983). An example is Elisabeth C. Stone's investigation of the Babylonian city of Nippur (Figure 10). By analyzing buildings in relation to the finds and the texts found in the houses, she has been able to reconstruct two different types of "neighborhoods" in the city. Stone points out that only a combined study of both artifacts and texts can show how the Mesopotamian cities were divided into these units, which were a form of loosely organized neighborhoods clearly linked to the surrounding countryside. Although the study concerns only a few blocks in the city of Nippur, Stone maintains that the loose urban structure she demonstrates can provide a general background to the loose political organization of the whole region into more or less independent city-states (Stone, 1987).

Literacy itself in Mesopotamia has also been critically scrutinized in recent years. Mogens Trolle Larsen argues that it is important to determine the changing function of writing through time (see Figure 35). He would tone down the significance of the first Sumerian writing, claiming that it should be seen more as a notation system, whereas a true literate culture cannot be detected until 1,500 years later in the Babylonian empire (Larsen, 1984, 1988; see also Nissen, 1986). At the

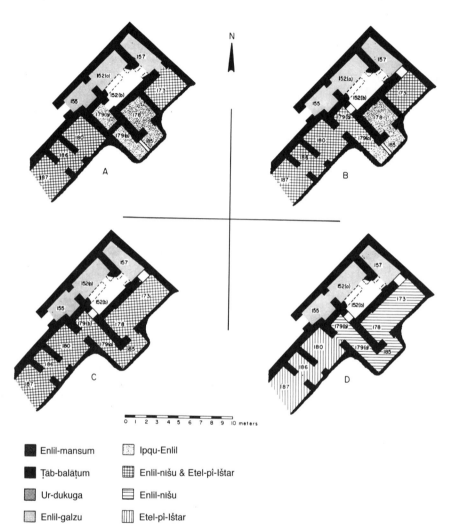

Figure 10. Ownership during four phases of settlement in the Babylonian city of Nippur (1742–1732 B.C.). The different shadings indicate properties owned by different persons (Stone, 1987, pl. 25, by courtesy of The Oriental Institute Museum). The figure is an example of the way an integrated analysis of urban settlement and cuneiform tablets found in the buildings can be used to reconstruct ownership in a Babylonian city.

same time, the long prehistory of writing in Mesopotamia has been studied in detail by Denise Schmandt-Besserat (1992). Through studies of small, standardized clay figures and "clay envelopes," she can follow the cognitive background over 5,000 years to the first writing.

With a stronger emphasis on the significance of archaeology, the relation of artifact to text has also been considered in recent decades. In a critical challenge to Mesopotamian archaeology, Maria de Jong Ellis (1983) asserts that the presence of text has prevented an independent development of method in archaeology. The lack of conscious integration of artifact and text has also meant that "well established knowledge concerning Mesopotamian history and social and economic institutions and their sequential development is a patchwork of interpretations based on studies using only part of the material that was available at the time of writing, certainly only a small part of what is available now" (Ellis, 1983:498). For Nicholas Postgate, archaeology can be used above all to check what the texts say about economic and social conditions, since "those who work with early administrative texts are continually aware that they contain detail on precisely those activities which can be reflected directly in the archaeological record" (Postgate, 1992:xxii). At the same time, he argues that one must compare a complete corpus of textual material with a complete body of archaeological material, to avoid the risk of selecting suitable correlating aspects (Postgate, 1990).

Defending the Faith: Biblical Archaeology

The fate of Palestine has been its role as a holy land for three monotheistic religions. In Judaism and Christianity in particular, Palestine represents a special sacred topography, since Jerusalem was seen as the center of the earth. For medieval Christian Europeans, Palestine was a distant place of pilgrimage, which came under Christian rule for a brief period in the twelfth and thirteenth centuries as a result of the crusades. After the Middle Ages, pilgrimages to Palestine became less common, and the area became as inaccessible as an imaginary landscape, being known only through the Bible. Although Europeans traveled in Palestine during the seventeenth and eighteenth centuries, it was not until around 1800 that systematic interest was aroused in the existing Palestine landscape and its ruins, which were then in a province of the Ottoman Empire (Silberman, 1982). Palestine, unlike Egypt and Mesopotamia, contained neither great monuments nor unknown writing systems, but through the Bible it was nevertheless a "historical" landscape. The biblical topography of Palestine was charted by the American Edward Robinson (1795–1863), who identified more than 200 biblical sites on trips in 1838 and 1852 (Moorey, 1991: 14 ff.).

With the religious debate in Europe and the finds of Mesopotamian parallels to the Old Testament texts, the idea arose that archaeology could serve as an ancillary science to biblical studies (Figure 11). Several societies for biblical archaeology were started in the 1870s, but the earliest professional archaeological excavations were not conducted until 1890, by the British Egyptologist William Flinders Petrie. As a direct consequence of the *Babel und Bibel* debate in Germany, major German excavations then got under way at Megiddo in 1903 and Jericho in 1907 (Moorey, 1991:1 ff.). The role of biblical archaeology at this stage was summed up by S. R. Driver in 1909, when he declared that archaeology "illustrates, supplements, confirms or corrects, statements or representations contained in the Bible" (cited from Moorey, 1991:44).

The interwar years saw the true breakthrough of biblical archaeology, and with it the emergence of two different traditions in the discipline. In Europe the subject had its base in Protestant Northern Europe and had a secular image. In the United States, the subject also emerged in Protestant settings, but there biblical archaeology had a more distinct religious profile (Moorey, 1991:54 ff.). Many active archaeologists were clergymen, and their work was expressly conceived as "the defense of the faith." This attitude is best seen in William Albright (1891–1971), who dominated biblical archaeology in America for decades. For him the purpose of archaeology was to historicize faith and ultimately to reconstruct "the route which our cultural ancestors travelled in order to reach Judaeo-Christian heights of spiritual insight and ethical monotheism" (cited from Moorey, 1991:72 f.).

For both the secular and the religious tradition, the Bible was the natural point of reference in all aspects of archaeological work. The objects of study were important sites in the Bible, such as Jericho, and public buildings, such as forts and city gates, that could be linked to the political history described in the Old Testament. The pottery chronologies were associated with the younger, "historical" parts of the Old Testament, and the finds were normally interpreted in biblical terms, such as the immigration of the Israelites (Moorey, 1991:54 ff.).

The postwar period has gradually seen several changes in biblical archaeology. With the foundation of the state of Israel in 1948, a new indigenous archaeology was created. Alongside European and American archaeologists, Israeli archaeologists now began to dig on a large scale. Although Israeli archaeology often has the Bible as its point of departure, the tradition is wholly secular and is often viewed with great mistrust by orthodox Jews. Moreover, the aim is different: to use archaeology to legitimize a controversial state formation (Moorey, 1991:87 ff.;

Figure 11. The Queen of Sheba visits King Solomon in Jerusalem. Biblical illustration from 1865 by Gustave Doré (1832–1883), here from the reproduction in a German translation of the bible by Martin Luther, *Die heilige Schrift*, 1877:552–553. Doré's illustrations, which were spread in translations of the Bible into many languages, were executed in a period when the very idea of biblical archaeology originated. The pictures represent a romantic version of the Middle East, although Doré received some inspiration from contemporary finds in Egypt and Mesopotamia. On the other hand, it is clear from the illustrations that Doré, like his contemporaries, knew little about the material culture of Palestine in the first millennium B.C. This lack of knowledge of realia was one reason for the emergence of biblical archaeology.

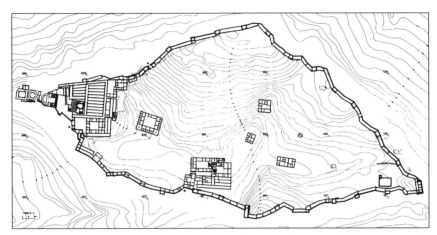

Figure 12. Restored plan of Masada at the end of the Herodian period (Netzer, 1991, plan 76, by courtesy of Israel Exploration Society). The excavations of Masada in 1963–1965 demonstrate the role of archaeology in the modern state of Israel. The excavations were carried out with the aid of the Israeli army and hundreds of volunteers from 28 countries. The results of the investigations were interpreted as an expression of the resistance of the Jews against Roman domination and were seen as a historical parallel to the state of siege in which modern Israel finds itself in the Middle East. For many years after the excavation, Israeli recruits swore their pledge of loyalty to the state of Israel on the rock of Masada, which also became the country's most popular tourist attraction. Yet Masada could also be used in criticism of modern Israel. In the early 1970s, several American critics described the uncompromising Israeli foreign policy as being dictated by a "Masada complex" (Silberman, 1989:87 f.).

Shay, 1989; Silberman, 1989:87 ff., 123 ff.). A good example of this nationalist trend is the excavation of Masada in 1963–1965 (Figure 12), under the leadership of Yigael Yadin (1917–1984). The rock of Masada was the last outpost of the Jewish revolt against the Romans in the seventies A.D., and according to the Jewish chronicler Josephus, the defenders of Masada committed collective suicide rather than surrender to the Romans. Archaeology seems to confirm Josephus's account in large measure, so Masada gave modern Israelis "a self-image of heroism and sacrifice when the country felt itself under the threat of imminent attack" (Silberman, 1989:100).

In European and American biblical archaeology, the basic tradition remained unchanged during the 1950s and the early 1960s, although the American tradition was sometimes perceived as religious fundamentalism by certain European archaeologists (Dever, 1990). The debate in this period largely concerned issues of excavation technique, pursued especially by Kathleen Kenyon (1906–1978). In the mid-1960s, however, landscape archaeology had its breakthrough in biblical archaeology as well, with large-scale Israeli and American survey projects. The concentration on the landscape as a whole, from the Neolithic until the Turkish period, has meant a successive break with historical geography and the biblical texts (Moorey, 1991:87 ff., 114 ff.).

As a result of this gradual emancipation from the texts, biblical archaeology in the United States in the 1970s has become a secular scholarship, influenced by new archaeology and completely independent of biblical studies. At the same time, the view of the Bible has changed in biblical scholarship, so that the texts are viewed today more as literature and less as historical texts. The secularization of biblical archaeology has in particular been pursued by William G. Dever since about 1970. He maintains that the subject should be called Syro-Palestinian archaeology, to release biblical archaeology from its strong methodological, religious, and political bonds with the Bible, and in particular the Old Testament (Dever, 1972; Moorey, 1991:135 ff.).

With the secularization of biblical archaeology, the relation between material culture and written sources has been discussed in a whole new way. H. J. Franken (1976) maintains that the "archaeology of a historical period is prehistory with one more problem," that is to say, the fundamental archaeological methods should not be affected by the presence of the texts. The text dependence of the earlier tradition he sees especially in the great interest in identifications, often very loosely based. Dever (1990) says that the relation between biblical texts and archaeology should be seen as a dialogue between two equal but very different parties. The Bible is above all a political and ideological document written and edited in the first century B.C. Archaeology, on the other hand, gives insight into economy, social conditions, and everyday life. Dever moreover stresses the significance of archaeology for tracing cultic practices not mentioned in biblical texts. Inscriptions, pictures, and temples suggest an extensive, noncentralized polytheism in Palestine, which can provide a religious background to the Old Testament texts about the prophets and their attacks on polytheism. Dever (1990:9) would therefore refuse to give the texts primacy of interpretation, since he sees the Bible as a "narrated artifact," to emphasize that both texts and artifacts need to be interpreted.

In a critique of the archaeological rhetoric in Syro-Palestine archaeology, however, J. Maxwell Miller (1991) argues that the Bible nevertheless serves as a reference point, since concepts and designations from the Bible are still used. "While it is theoretically possible to write a history of early Israel without relying on the Hebrew Bible, the result would be a very thin volume indeed and would have little in common with current discussion" (Miller, 1991:101).

ASIA

In Asia, to the east of the Middle East, most countries have historical archaeologies, but they are rarely distinguished as specialized subjects as in Europe and the Middle East. Instead, there are elements of historical archaeology in most regionally oriented archaeologies.

As in the Middle East, archaeology in Asia is of Western origin, closely associated with the European and later the American dominance of the continent. And as in the Middle East, a knowledge of the past, partly through archaeology, became a way for the great colonial powers to legitimize their political dominance (Said, 1978).

From a European point of view, the continent was less well known than the Middle East, although, for example, India was mentioned by ancient geographers and historians. It was not until the expansion of European trade in the sixteenth century that Europe came into direct contact with the Asian coasts. Europeans did not really encounter Asia and its often unknown cultures in earnest until the colonial conquests of the late eighteenth and the nineteenth centuries. In this encounter we can detect the first efforts at archaeological work. With the exception of China and Japan, the Asian archaeological scene was long dominated by the colonial powers, but archaeology today often shows a clear indigenous profile and basis.

There are many interesting historical archaeologies in Asia (e.g., Higham, 1989; Nelson, 1993; Smith and Watson, 1979). In this context, however, I shall deal only with three archaeological traditions: Indian, Chinese, and Japanese archaeology. What is particularly interesting about these subjects is that they have long been dominated by indigenous archaeologists, despite the Western origin of archaeology. Yet the emergence of these native traditions has led to archaeologies of radically different character in the three countries.

Exotic and Familiar: Indian Archaeology

Indian archaeology is closely connected to the colonial history of the region. European interest in the history and religions of India was aroused with the expansion of European trade from around 1500. European travelers began in the sixteenth century to describe Indian monuments, especially temples and caves. Their attitude was often ambivalent, vacillating between admiration and scorn for the alien architecture and art. The Greek influences in Gandhara were detected at an early stage, however (Chakrabarti, 1988:1 ff.). Knowledge of India was mainly conveyed to Europe in the sixteenth and seventeenth centuries by Jesuit missionaries. Since they saw the Hindu belief in a "supreme being" as a parallel to the Christian God, their mission was not geared to abolishing Hinduism but rather to bringing it to Christian "perfection" (Halbfass, 1981:65 ff.).

The Jesuit missionaries' descriptions of India were subsequently integrated in European debate and culture. The "Christian" character of Hinduism was used as an argument for deism, that is, the idea of a universal rational religion, beyond culture-specific ritual. During the Enlightenment the culture and religions of India seemed an alternative to the European Judeo-Christian tradition. Voltaire claimed that India had the oldest culture and the most original religion in the world (Halbfass, 1981:70 ff.). The romantics instead cherished the notion that India represented the childhood, innocence, and purity of humanity. The relationship of Sanskrit to the languages of Europe had been observed in the sixteenth century, but when William Jones (1746–1794) in 1786 formulated the idea that this relationship was due to a common historical origin, an almost mystic community between Europe and India was established. This was further strengthened by the fact that several of the romantic poets were influenced by Indian literature, which became accessible in Europe for the first time via translations in the early nineteenth century. The linguistic bond between Europe and India meant that studies of Sanskrit language and literature became an important part of the study of what came to be called from the start of the nineteenth century "Indo-European" or "Indo-Germanic" (Bernal, 1987:224 ff.; Chakrabarti, 1988:15 ff.; Halbfass, 1981:86 ff.).

Systematic studies in India of the country's past began at the end of the eighteenth century, in connection with British colonial expansion. An important factor in the study of India's early history was that there was no secular historiography before the seventh or eighth century A.D. Earlier historical writing was embedded in religious literature, which was written on the basis of an almost infinite concept of time. Many Europeans viewed the early historiography as "timeless" (Paddayya, 1995), so the study of early Indian literature was soon linked to investigations of material remains with the aim of reconstructing India's history. This was obvious in the Asiatic Society, founded by William Jones in 1784, two years before he formulated his famous thesis. The aim was to study the history of India through antiquities, art, and literature (Chakrabarti, 1988:15 ff.).

In the first half of the nineteenth century, the study of India's past was connected with individuals acting on their own initiative. An illustrative example is the Scottish indigo cultivator James Fergusson, who established the first chronological and geographic framework for Indian monumental architecture during his many years of travels in the 1830s and 1840s. Other important work was carried out by James Prinsep (1799–1840), who succeeded in deciphering the two oldest Indian writing systems—Brahmi and Kharosthi—in the 1830s. The decipherment gave scholars access to the texts of tens of thousands of inscriptions from the period 250 B.C. to A.D. 1200, not least the famous edicts of Aśoka from the first Indian empire in the third century B.C. (Chakrabarti, 1988:32 ff., 98 ff.).

An important turning point in the view of India, and simultaneously in the study of India's past, occurred in connection with the Sepoy Mutiny of 1857–1858. The formerly positive image of ancient India was then replaced by a more negative one, in which India was increasingly viewed as a passive culture receiving influences from outside (Chakrabarti, 1988:21), and dominated by "monstrous superstition" (Halbfass, 1981:84). The revolt against British rule in India led to the transfer of colonial administration from the East India Company to the British state, which also meant that archaeological work acquired a firmer organization.

The state-financed Indian Archaeological Survey, under the direction of Alexander Cunningham (1814–1893), came into existence not long after, in 1861. The work was concentrated on the survey of historic sites, monuments, inscriptions, and sculptures in northern and central India (Figure 13). By combining ruined cities and monuments with inscriptions, Indian literature, Greek sources, and especially two travel accounts by Buddhist pilgrims from China, Cunningham succeeded for the first time in establishing a historical and geographic framework for the early history of India (Chakrabarti, 1988:48 ff.). He was particularly interested in Buddhist remains, since they represented the oldest "historical" epoch in India, and since they could show that Hinduism was not the only Indian religion; it was hoped that this would further the Christian mission. Dilip Chakrabarti (1988:44) expressly maintains that "Cunningham was trying to justify systematic archaeological exploration of India on the grounds that it would politically help British rule in India and would lead to an easier acceptance of Christianity in the country."

In the 1880s the surveys were extended to cover southern and western India and to include monuments from all historical periods. The monuments were protected by law, and the most important build-

Barrow D   Barrow E   Barrow F   Barrow G   Barrow H   Pippal Tree

Figure 13. Lion-crowned column at Lauriya Nandangarh in 1862 (Cunningham, 1871: 68–69). The column bears one of the edicts of the Emperor Aśoka (ca. 272–231 B.C.), in which he exhorts the population to live in accordance with a more or less explicitly Buddhist ethic (dharma). The picture illustrates how British archaeologists were quick to compile inventories of monuments and inscriptions in order to write the early history of India. Since there was no contemporary secular historiography, the maximum extent of the Mauryan Empire during the third century can only be revealed by charting Aśoka's edicts.

ings began to be restored (Chakrabarti, 1988:94 ff.). As in Europe, studies of historical architecture provided a model for a local historicizing architectural style: Indo-Saracen. In both Europe and India, then, there was a close bond between architectural history and the development of a historicizing idiom. The Indo-Saracen style was an expression that the British, after the Sepoy Mutiny of 1857–1858, viewed themselves as the political heirs of the overthrown Mogul dynasty and expressed this heritage by means of symbolic associations with India's past (Davies, 1985; Nilsson, 1968).

As in Europe, the historicizing styles were rejected at the start of the twentieth century. Also as in Europe, concern for the historical monuments changed from practical applications to an interest in their significance in the past. In the 1920s the Indian philosopher and art historian Ananda K. Coomaraswany (1877–1947) began studies of the meaning of Indian temple architecture. By comparisons between early texts and surviving buildings, he reconstructed the cosmological meaning of several temples. In these studies he also searched for a timeless metaphysics that could be valid for all preindustrial societies in the world, and which could serve as a critique of modern Europe (Halbfass, 1981:288 ff.; Paddayya, 1995).

Archaeological excavations proper did not begin until 1902, when John Marshall became the head of the Indian Archaeological Survey. He was a classical archaeologist who introduced up-to-date excavation techniques to the first generation of indigenous Indian archaeologists. The excavations initially concentrated on the oldest "historical" cities in the Ganges basin, which are mentioned both in Greek sources and in early Indian literature (Chakrabarti, 1988:120 ff.). Through Alexander the Great's campaign, which affected some of these cities, there was a link between India and the classical world of the Mediterranean. From an Indian point of view, too, the cities were important since the historical Buddha, Prince Siddhartha Gautama, was active in this area, probably in the fourth century B.C. Moreover, these cities represented the very core and starting point of the first "historical" Indian state, the Mauryan empire, in the third century B.C.

With the sensational discovery of the Indus culture in the 1920s, however, archaeological interest shifted to the much older cities in the Indus basin (Chakrabarti, 1988:156 ff.). The unearthing of the Indus culture raised several fundamental questions in Indian archaeology, some of which have not yet been answered. The script, which consists of about 250 basic signs and 50 auxiliary characters, has still not been deciphered. And the relations between the urban culture of the Indus valley and the thousand-year younger urban culture in the Ganges valley is unclear. This applies not least to the religious ritual and iconography of the Indus culture in relation to Hinduism and Vedic literature (see Allchin and Allchin, 1982:212 ff.; Conningham, 1995). During the economic crisis of the 1930s, archaeological activity was at a low level, but in the 1940s Mortimer Wheeler renewed excavation techniques with a number of new investigations. An important example is his excavation of the trading site of Arikamedu in southern India, which had links with the Roman Empire (Chakrabarti, 1988:173).

Indigenous Indian archaeologists began to be trained from the start of the twentieth century, but until independence in 1947, Indian archaeology was hampered by a Eurocentric view of India as a passive culture heavily influenced by religion and spiritualism. The British archaeologists who dominated the subject were primarily interested in detecting foreign influence on Indian culture, art, and architecture. Above all, they stressed the Greek, Roman, and Iranian impulses (Sharma, 1987: 1). A typical expression of this "orientalist" image among British archaeologists is Stuart Piggott's (1945:1 f.) perception of India: "We do not find, and should not look for an inherent element of progress in Indian history—no organic evolution of institutions to changing human needs, no development of material culture nor the gradual spread of higher standards of living to a constantly increasing proportion of the inhabitants."

With Indian independence and partition in 1947, most of the cities of the Indus culture ended up in Pakistan, where continued archaeological work has been dominated by foreign expeditions. In present-day India, on the other hand, an indigenous archaeology has developed considerably since independence (Allchin and Allchin, 1982:1 ff.). Because very little of the Indus culture is accessible in India, Indian archaeology has once again turned to the early historical cities in the Ganges basin (Erdosy, 1988:1 ff.). The focus is on the early Hindu and Buddhist civilizations, especially from the period 500 B.C. to A.D. 500, with their cities, temples, monasteries, stupas, and graves. In particular, the sites and monuments that are known from Indian literature have been excavated. In contrast, less attention has been devoted to later periods and the Dravidian south of India (Deshpande, 1967; Sharma, 1987:1 ff.).

In postcolonial archaeology, different basic views of India's past have been at the center of debate. Indian historians were early in criticizing the orientalist image of India, and in archaeology S. C. Malik (1975) questioned the stereotyped view of India as a passive, unchanging culture. He argued that the traditional perception could be changed by applying European social theory to India's past. Another reaction to orientalism has been a constant emphasis on the indigenous origin of phenomena in ancient India, but this attitude has also been criticized for being "chauvinistic and parochial" (Sharma, 1990:4). The issue of native versus foreign features has in recent years been taken to extremes with the Hindu attacks on Muslim monuments. The politically explosive situation after the demolition of a mosque in Ayodhya has divided Indian society, and Indian archaeology, into different camps with very different views of history (Colley, 1995; Mandal, 1993; Paddayya, 1995; Rao, 1994).

The perspective in postcolonial archaeology in India has long been characterized by a tradition of "cultural history," concentrating on chronological and spatial groupings of objects. In the 1970s and 1980s, however, economic and social questions have come into focus, partly through collaboration with economic and social historians, partly through the influence of processual archaeology (Paddayya, 1995). One expression of this perspective is landscape archaeology, which had its breakthrough in the 1980s, when large-scale surveys were conducted around the oldest historical cities in the Ganges valley (Lal, 1984).

In recent years there has been a concern with the meaning of artifacts, partly as a result of a resumption of Coomaraswany's interwar studies of the meaning of temple architecture. Besides individual buildings, complete cities with streets and monuments have also been studied. By means of comparisons between texts and the form of the cities, several scholars have emphasized the cosmological meaning of town plans from diverse periods in India's history (see Figure 14). These cosmological studies may be exemplified by the large-scale investigations of Vijayanagara, which was the capital of the last Hindu empire in India, circa 1340–1700 (Fritz and Mitchell, 1987).

Postprocessual archaeology, with its emphasis on the ambiguity of meaning, has also been seen as a possible way to detach South Asian archaeology from its theoretical dependence on Europe and the United States (Manatunga, 1994; Paddayya, 1990:50, 1995). Paddayya (1995) points out that ambiguity is prominent in a long tradition of Indian textual interpretation, and that there are indigenous concepts for time, place, and cause and effect, which could be an Indian contribution to the predominantly Western theoretical debate in archaeology. This possibility of linking Indian and European philosophy follows a long Indian tradition of "comparative philosophy" (Halbfass, 1981:289).

Both for the politically sensitive issues and for the methodological problems in Indian archaeology, the view of the relation between artifact and text is significant. The controversial question of native versus foreign in India's past concerns in large measure the view of the Aryans and the simplified question of whether they invaded or not. At the center of this debate is the Vedic literature and the question of whether and how it can be related to archaeology. Many attempts have been made to correlate the texts with material culture, but Chakrabarti (1984) believes that the whole approach—trying to pinpoint the "timeless" Veda in time and place with the aid of archaeology—is misguided. For him, archaeology and literature cannot be compared here, since the archaeological image of protohistorical India differs from Vedic literature, which is a conglomerate of texts from very different periods and contexts.

In connection with the early historical cities in the Ganges basin, the *Arthasastra*, a work on statecraft probably dating from around 300 B.C., has been compared with archaeology (Figure 14). The cities have been correlated with the ideal city plan in the text, and survey results have been compared with descriptions of the ideal society (Allchin, 1995; Erdosy, 1988; cf. Wheatley, 1971:414 ff.). Texts have also played a major

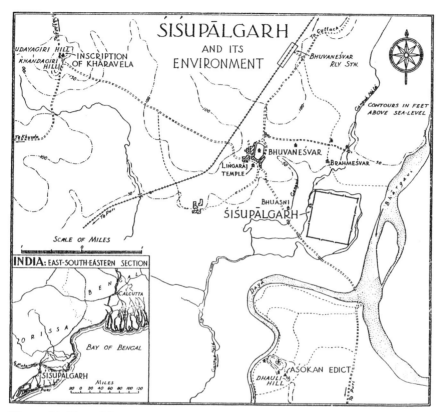

Figure 14. Plan of the remains of the city of Sisupalgarh and its immediate surroundings (Allchin, 1995, fig. 11.7, by courtesy of Archaeological Survey of India). The city was fortified in the latter half of the third century B.C., after Emperor Aśoka's conquest of eastern India. Comparisons with *Arthasastra*, a work on statecraft probably from around 300 B.C., show that the regular layout of the city and the distribution of functions inside and outside the city walls come very close to the ideal of the Indian city described in the text (Allchin, 1995; Wheatley, 1971:414 ff.).

role in the interpretation of the cosmological meaning of Vijayanagara (Fritz and Mitchell, 1987).

The question of the relation between material culture and writing has also been considered from a more theoretical point of view. The historian R. S. Sharma (1990:5 ff.) argues that the definition of historical archaeology as "text-aided archaeology" confines the perspectives. He would also like to see "archaeology-aided texts." George Erdosy (1988) maintains that the dominance of texts in historical archaeology is due to the fact that we live in a textual world that requires us to translate the objects studied by archaeology into text if they are to have any meaning. Since all meaning in archaeology comes from analogies, he pleads that archaeological analyses should be performed as independently as possible before analogies, whether historical or ethnographical, are drawn.

History as Morality: Chinese Archaeology

Chinese archaeology has been created in a unique encounter between a European science and an indigenous antiquarian tradition. The native scholarly tradition ultimately goes back to the fifth century B.C. with the redaction of the oldest historical annals, and to the second century B.C. with the composition of the first Chinese universal history. Above all, antiquarian studies derive from the Sung dynasty (960–1279), when studies of ancient artifacts and inscriptions were begun as a means to complement the classical texts. Through the centuries, the amassed historical and antiquarian knowledge became the framework of traditional Chinese learning. This was primarily viewed as moral knowledge, since it was considered necessary for ruling (Chang, 1981; von Falkenhausen, 1993). In the second century B.C., the author of the first Chinese universal history, Sima Qian, claimed that "events of the past, if not forgotten, are teachings about the future" (cited from Chang, 1986:6). In modern times, Mao transformed this ideal of learning into the slogan "Let the past serve the present" (Capon, 1977).

China was known in Europe from the thirteenth century, but it was not possible to have any profound knowledge of Chinese culture until the early sixteenth century, in conjunction with the expansion of European trade. As in India, information about China was primarily transmitted by Jesuit missionaries, and as in the case of India, their information about Chinese religion was used as arguments for deism. The observations of the Jesuit Matteo Ricci became particularly important in the European debate, since he described the Chinese state as borne by mandarins with a thorough knowledge of Chinese philosophy. Philosophers such as Spinoza, Leibniz, and Voltaire thus viewed China as a utopian ideal, a state governed by philosophers (Dawson, 1967: 35 ff.; Malcus, 1970:27 ff.). China's superiority was also thought to be confirmed in material terms by Chinese luxury products, such as silk and porcelain, which were spread over Europe by the East Indian companies. The positive image of China in Europe may also be seen in the way that Chinese motifs were integrated in European architecture and art, especially in the chinoiserie of the rococo (Dawson, 1967: 106 ff.; Malcus, 1970:29 ff.).

With romanticism and the birth of the idea of evolution, the European view of China became more negative. The country was increasingly seen as stagnant and lacking history, and at the same time Protestant missionaries described the "spiritual destitution" in China (Malcus, 1970:30 ff.). This negative attitude was reinforced by the Opium War of 1839–1842 and the Arrow War of 1856–1860. As a result of these wars, which forced China to open its boundaries to Europeans and Americans, foreigners also began to come across traces of the early history of China. With the construction of railways in the late nineteenth and early twentieth centuries especially, many remains of ancient Chinese history were discovered. A great deal of the Chinese collections in Western museums come from graves uncovered by the railway navvies, in many cases after downright grave robbing (Capon, 1977).

Some foreign archaeologists and geologists carried on archaeological work during the first decades of the twentieth century, but professional archaeology was not established in China until the 1920s, when Chinese archaeologists who had received their training in Europe or the United States started working. The introduction of archaeology was a part of the reformist "Fourth of May Movement," which tried from 1919 onward to replace traditional Chinese learning with Western science. Under the leadership of Li Ji (1896–1979), archaeology was initially concentrated on sites associated with the early history of China. The excavations of Yinxu, the last capital of the Shang dynasty, in 1928–1937, were to be particularly influential for archaeology in China. Since the 1890s, the site has been known for finds of tens of thousands of oracle bones from about 1200–1050 B.C., and through these bones the Shang dynasty has become the oldest "historical" period in China (Figure 15). Yet although archaeology was introduced as a Western alternative, Chinese archaeology gradually became linked to indigenous scholarship. Despite attempts by Li Ji to establish archaeological typologies, artifacts were classified according to the antiquarian tradition founded in the Sung dynasty, which means that today there is still no generally accepted typological nomenclature for archaeological finds from China (Trigger, 1989:174 ff.; von Falkenhausen, 1993).

The creation of the People's Republic of China in 1949 was the start of a new phase in Chinese archaeology. All plundering stopped, and archaeology became an important state activity, with two main aims. One was to emphasize the long history of the country, its great culture and national unity, by means of excavations, finds, and exhibitions. The other was to confirm Marxist historiography and thus legitimate the power of the Communist Party. Because of these patriotic and ideological aims, archaeology paradoxically became more closely linked

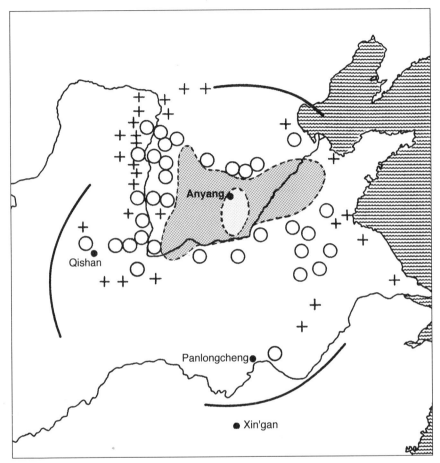

Figure 15. Reconstruction of the late Shang state (ca. 1200–1050 B.C.) and its immediate surroundings based on archaeology and inscriptions on oracle bones (Barnes, 1993, fig. 60, by courtesy of Thames and Hudson). Ever since the establishment of archaeology in China, there has been great interest in the Shang dynasty as the oldest historical period in the country. This map is the latest attempt to visualize the late Shang state. At the center is the capital, Anyang (Yinxu), with the "inner domains" marked by densely grouped sites with remains of the Shang culture. Outside this area are the "outer domains," which in turn are surrounded by friendly rulers (rings) and ultimately by hostile lords (crosses), according to the oracle bone inscriptions. The late Shang state cannot be perceived as a political territory with fixed borders, but should instead be viewed as a political organization based on shifting alliances (Barnes, 1993:131 ff.).

to the antiquarian tradition than it had been in the interwar years (von Falkenhausen, 1993).

Archaeological work was concentrated on the origin of China in the valley of the Yellow River. The excavations have mainly concerned cities, temples, palaces, and rich graves from the Shang dynasty until the Zhou dynasty (1700–221 B.C.), along with occasional graves from the Han and Tang dynasties (206 B.C.–A.D. 906). Interpretations of the finds have often stressed "centralism," the spread of phenomena from a center. As Lothar von Falkenhausen (1993) has pointed out, nowhere else in the world has postwar archaeology been as geared to the elite as in China. It is also thought provoking that there is no landscape archaeology in China.

The clearest example of the close association between archaeology and politics is the excavation of the famous terracotta army at the grave of Emperor Qin outside Xian (Capon, 1977; von Falkenhausen, 1993). The excavations began in 1974 as part of an anti-Confucian campaign. Parallels were drawn between Mao and Qin, since both were seen as the first rulers in new phases of China's history, and both were opposed to Confucius's aristocratic worldview. The realistically portrayed terracotta soldiers were also highlighted in the political debate, where they were held up as examples for contemporary social realism in art (Figure 16).

The introduction of capitalism in China since 1987 has meant great changes for archaeology. Many historical cities are now being rapidly rebuilt (Ruan, 1993), and large-scale plunder of graves has resumed, in several cases with the aid of local authorities. At the same time, archaeology has lost its role as a creator of national identity, so state resources have been drastically reduced. Archaeology is in financial crisis, but there has simultaneously been a certain degree of liberation from Marxist historiography. In souther China in particular, archaeology has begun to follow an anthropological trend. In some cases the traditional dependence on texts has been questioned, and Marxism has been criticized for being a clumsy foreign dogma (von Falkenhausen, 1993).

With the incorporation of archaeology in classical Chinese historiography, historical archaeology in China has primarily been viewed as a complement and a corrective to written sources. This view has also been reinforced by the fact that many excavations have produced new texts. Over 150,000 oracle bones have been found to date (Chen, 1989), and these texts give detailed insight into religious and political conditions in the late Shang period (ca. 1250–1050 B.C.). In addition, there are cast inscriptions of varying length on many bronze vessels from the Shang and Zhou dynasties. Younger graves have also yielded texts

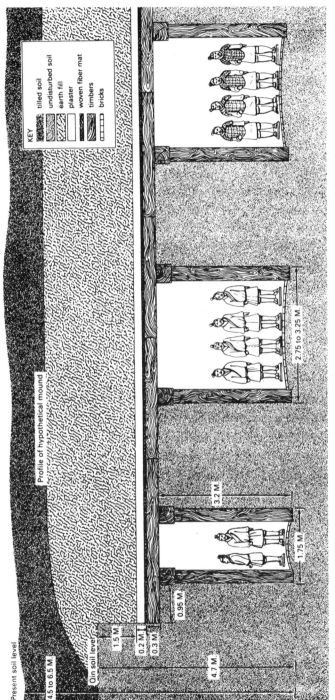

Figure 16. Trenches with terracotta soldiers at the grave of Emperor Qin near Xian (based on Fong, 1980, fig. 121, Metropolitan Museum of Art). The terracotta army, which is estimated to consist of more than 3,200 soldiers, was created as a form of guard around the huge burial mound of the emperor. The find of the terracotta soldiers is one of the most sensational archaeological discoveries of recent decades. Yet the excavations, which began in 1974, had clear political undertones, as a way to provide historical associations for Mao and his policies during the Cultural Revolution (von Falkenhausen, 1993).

written on bamboo and silk. These inscriptions from archaeological contexts have long been used in studies of the history of language and writing. The inscriptions on the oracle bones in particular have been important for an understanding of the pictorial and semantic background to many Chinese characters (Lindqvist, 1989).

From an archaeological point of view, however, both the finds of inscriptions and the long textual domination of learning in China have meant that the discussion of artifacts and texts has often stopped at questions of identification and classification. Hsia (1986) has tried to renew this debate by juxtaposing designations for jade objects in the antiquarian tradition of the Sung dynasty with modern archaeological typologies. Yet there are also examples of archaeologists and anthropologists—both inside and outside China—who have dealt with questions going beyond basic classification. Paul Wheatley (1971:122 ff.) perceives writing as an expression of the ruling elite and its actions, values, and attitudes to other groups in society. By comparing early texts with excavated cities, he is able to show how the great Chinese cities and their monuments are laid out according to ideal plans that more or less clearly reflect the cosmological outlook of the elite. He finds the same association between texts, monuments, and town plans in many other "early" civilizations over the world. K.-C. Chang has pleaded in several studies (1976, 1983, 1986, 1989) for an interdisciplinary study of the early history of China, so as to create an alternative to the stereotyped Western image of China as an unchanging, stagnant country. By combining artifacts, texts, and pictures, he is able to demonstrate, for example, a little-known element of shamanism in the Shang state and to suggest how the use of baskets and vessels of bronze, clay, and wood can be correlated with the doctrine of the five elements (Chang, 1976: 115 ff., 1983:44 ff.).

The significance of the archaeological context of writing has been touched upon in connection with a burial place from the Shang dynasty at Yinxu. By comparing the spatial placing of the graves with the clan marks on the grave vessels, Noël Barnard (1986) is able to show complicated kinship relations between the buried persons. The contrast between artifact and text has also been emphasized as an important dimension. By comparing the political geography of the Shang state with the archaeologically defined "Shang culture," David N. Keightley (1983:523 ff.) paints a new picture of the Shang state that would not be obvious from either artifacts or texts on their own.

The very principle of the textual dependence of Chinese archaeology has been criticized by several scholars. Xia Nai (1910–1985) tried to make writing and material culture more equal by seeing archaeology as a text of its own, to be critically compared with writing (von Falkenhausen, 1993). R. W. Bagley (1992) maintains that the texts constitute a double danger for archaeologists, since they restrict both seeking and seeing. Finally, von Falkenhausen (1993:847) points out that the problems in Chinese archaeology "are common to archaeology of historical periods world-wide; they are, however, magnified, in the Chinese case, by the vastness of the historical record and by the centrality of the historical texts in the culture."

In Search of a New Identity: Japanese Archaeology

As in China, archaeology in Japan has been created in the encounter between a tradition of indigenous learning and a European science. The two oldest-known chronicles of Japan (*Kojiki* and *Nihon Shoki*) were written down in the early eighth century A.D. (see Figure 17). The twelfth century saw an early interest in old traditions and artifacts, such as palaces, clothes, and weapons, as Japan was rapidly transformed into a society dominated by many warlords. However, antiquarian studies of artifacts from the past did not begin on any scale until the Tokugawa period (1603–1867) (Barnes, 1990; 1993:28 f.). Unlike China, this was a case of a native antiquarian tradition created partly through European influence.

With the expansion of European trade, the first Europeans visited Japan in 1543. In the following hundred years, extensive contacts and exchanges between Europe and Japan developed. As in India and China, much of the information about Japan was transmitted by Jesuit missionaries. In reaction to the increasing European influence, however, the Japanese rulers decided to isolate the country from the outside world in 1639. Contacts with Europe were maintained through a single Dutch trading station, and it was via this place that European impulses were communicated to Japan. Examples are the European linear perspective that was accepted in Japanese art in the mid-eighteenth century, and the European science—known as *rangaku*—that was introduced from the latter half of the seventeenth century. Special attention was devoted to the significance of systematic collection and depiction in the natural sciences (Croissant and Ledderose, 1993).

Indigenous antiquarian studies were influenced by Western natural science, since it could be combined with the contemporary neo-Confucianism and its rationalist emphasis on studying things in this world. During the Tokugawa period, artifacts from the past began to be studied from two perspectives. In "the Naturalist and Rock Fondler traditions" collections were built up of rocks, fossils, and stone tools, which are perceived today as prehistoric, while "the Court Etiquette and Neo-Confucianist traditions" were concerned with historical monuments and texts. The first archaeological excavation was carried out from the latter perspective in 1692, and the first half of the nineteenth century saw surveys of burial mounds that could be associated with historical persons. These antiquarian studies became part of Japanese learning during the centuries when Japan was largely a closed country (Barnes, 1990; Bleed, 1986).

The turning point in Japan's relations with the rest of the world did not come until 1854, when the country was forced to open its boundaries under the threat of war. Europeans conceived a keen interest in the newly opened Japan; Japanese woodcuts, for example, played a decisive role in the emergence of impressionism and Jugend (Budde, 1993). Japan began a deliberate modernization of its society, especially after the Meiji Restoration in 1868. The country was rapidly industrialized, and by the start of the twentieth century it was already one of the world's great powers. Archaeology was introduced as a part of this huge wave of modernization, with a national museum being founded in 1871 in the new capital of Edo/Tokyo (Kidder, 1977). During the first decades, it was mainly foreign archaeologists who worked in Japan, but in the 1890s an indigenous Japanese archaeology was established, with close ties to the former traditional learning. Chronology was bound by the empire's official view of history, which was mainly based on the two early-eighth-century chronicles (Figure 17). There was thus no room for any "prehistory" before the testimony of the written word. Paradoxically, early Japanese archaeology concentrated on what is today viewed as the Stone Age, so there was great debate about how the artifacts should be interpreted in relation to traditional Japanese chronology. The frames for interpretation were gradually restricted by the harsher political climate in the 1920s and 1930s (Barnes, 1990; Junko, 1989; Trigger, 1989:177 ff.).

Japan's military and political defeat in 1945 meant that archaeology completely changed character after the war. The subject enjoys broad popularity today, probably because studies of Japan's past have filled the ideological vacuum that arose after the defeat. Archaeology has played an active role in *Nihon-bunka-ron*, the attempts to explain Japan's economic success since the 1970s on the basis of nationalist ideas of a superior, homogeneous, unchanging Japanese culture (Trigger, 1989:179; Tsude, 1995:302 f.).

Although Japanese archaeology before the Second World War was seen as "historical," it was only after the war that archaeology also began to concern itself with historical periods in the strict sense. It was not until the 1960s that historical archaeology on any scale was established. As in European medieval archaeology, however, the subject

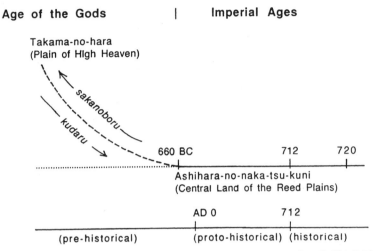

Age of the Gods | **Imperial Ages**

Figure 17. Traditional chronology in the early Japanese chronicles *Kojiki* (712) and *Nihon Shoki* (720) compared with archaeological periodization (Barnes, 1990, fig. 1, by courtesy of Gina Barnes and Antiquity Publications). The short chronology and the divine origin of history, which dominated the Japanese perception of history until the end of the Second World War, were difficult to combine with Western archaeology's secularized view of history and almost infinite temporal depth. In addition, it was methodologically difficult to reconcile the two historical perspectives, since older periods were viewed as "higher up" in traditional Japanese chronology, whereas older periods are normally "further down" in the stratigraphic way of thinking in archaeology (Barnes, 1990).

partly derives from earlier studies of architectural history. Studies of the old imperial city of Nara began in the 1890s, to find out whether the monuments had been rebuilt after the foundation of the city. In the 1920s and 1930s, some palaces and temples were also excavated (Tanaka, 1987; Yamamoto, 1986).

Historical archaeology today is chiefly concerned with the Japanese state formation and the early history of the united country, that is, the period A.D. 200–800, although later periods are sometimes studied too (Barnes, 1993:267 ff.). The main archaeological issues regarding the state formation are the demographic, political, and religious background to the establishment of the state. An important but politically sensitive question is whether the emergence of the Japanese state was connected to a conquest from Korea. A central position in this discussion is occupied by the aristocratic "keyhole-shaped" grave mounds (see Figure 31), which can be partly linked to persons in early Japanese history (Barnes, 1988; Tsude, 1990).

In relation to the united Japan, the religious, political, and administrative foundations of the state have been a particular subject of study. Great attention has been devoted to temples, partly because Buddhism was introduced as a form of state religion when the country was politically unified (Figure 18). Archaeological excavations have shown that the structure of the temple areas and the individual monuments follow different religious traditions in their function and style. Links between different temples and between temples and their properties have also been detected as a result of studies of roof tiles (Yamamoto, 1986).

Alongside temples, considerable attention has been devoted to the early imperial cities and their palaces, and to early administrative centers in the different provinces in the country. Archaeology has been able to show how the imperial cities and palaces were built according to Chinese ideals of town planning, and how the central places in the provinces were smaller versions of the imperial palaces. The early Japanese state thus manifested itself as a spacious symmetrical order that was present throughout the country, and that deviated noticeably from the normal houses with sunken floors, which were built closely together in irregular patterns (Tanaka, 1987). By investigations in temples and palaces, archaeological excavations have also yielded new

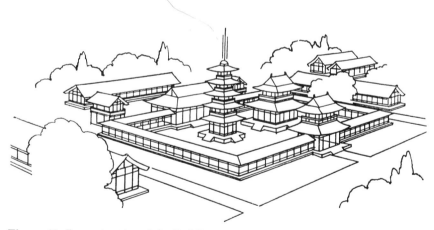

Figure 18. Reconstruction of the Buddhist Horyuji temple from the eighth century at the imperial city of Nara (Paine and Soper, 1955, fig. 3A, by courtesy of Yale University Press Pelican History of Art). The reconstruction is based on still surviving wooden buildings and archaeological investigations. Although Buddhism was not formally a state religion, it was furthered by the political elite who created the Japanese state. Buddhism and its institutions are thus important aspects of Japanese state formation in the seventh and eighth centuries (cf. Figure 31).

written documents. Since 1961, over 100,000 *mokkan*, or wooden tablets with writing, have been found at about 50 sites. These wooden tablets have allowed a closer study of early Japanese administration from the end of the sixth century to the end of the eighth century (Tanaka, 1987; Yamamoto, 1986).

Since the 1970s, Japanese archaeology has been dominated by rescue excavations; in 1990 alone there were about 25,800 excavations. Archaeology has become so big and difficult to survey that Japanese archaeologists have been described as "victims of their own success" (cited from Tsude, 1995:299). The idea of the homogeneity and continuity of Japanese culture has moreover meant that the interpretation of the past has been viewed as relatively unproblematic. This has led to a highly empirical archaeology with rather little debate about theory and method (Barnes, 1984; Junko, 1989; Matsumoto et al., 1994; Tsude, 1995). Historical archaeology likewise has an empirical profile, with a debate that is mostly confined to identification and classification. Gina L. Barnes (1984) thinks that there is a tendency to harmonize material culture with writing, in what she calls "the matching game." She finds this tendency particularly unfortunate, since the ambiguity of the written Japanese language easily invites one to search for different types of agreement. In her critique, Barnes therefore calls for introductory independent pattern analyses of artifacts and of texts, instead of working with isolated details, which can be more easily harmonized. Instead of just illustrating texts, archaeology could also be used to elaborate and challenge written sources (Barnes, 1993:16 ff.). Barnes herself has tried to renew the discussion of Japanese state formation by studying demography and settlement in early historical Japan (Barnes, 1988).

Historical Archaeologies in Africa and America | 4

AFRICA

In Africa south of the Sahara, archaeology and historical archaeology are very unevenly spread. Some countries have virtually no archaeology, whereas others, such as Nigeria, Kenya, and South Africa, have extensive archaeological activity. As in the Middle East and Asia, archaeology is of European origin. Before the expansion of trade in the late fifteenth century, Africa south of the Sahara was largely unknown to Europeans. Even when European trade reached Africa, European merchants usually stayed in their fortified trading stations along the African coasts. Although they rarely left these coastal areas, they had a constant effect on the African interior through their trade, especially in connection with the destructive slave trade from the start of the sixteenth century until the first half of the nineteenth century (see Wolf, 1982:195 ff.). It was only with the colonial conquest and division of Africa at the end of the nineteenth century that Europeans began to penetrate the whole continent. It was in this colonial encounter with "primitive" Africa that Europeans began to collect artifacts and excavate. African archaeology is still dominated today by Europeans and Americans, although there are indigenous archaeologists in some countries.

Although Africa is a huge continent, I shall present African archaeology as a whole, but taking examples from different parts of the continent. The reason for this summary treatment is that Africa's shared colonial history has given archaeology similar conditions and character over the whole continent. Moreover, African archaeology has been presented as a unity in several contexts in the last few years (cf. Ki-Zerbo, 1981; Robertshaw, 1990a; Schmidt, 1983; Shaw, 1989; Shaw et al., 1993).

Heart of Darkness: African Archaeology

Archaeology in Africa is intimately associated with colonialism and its ideology. The coasts of Africa were first described by the Portuguese at the end of the fifteenth century, when the first African artifacts were also brought to royal art collections in Europe (Vansina, 1984:21 ff.). With trade, and above all the slave trade, more and more information about Africa came to Europe. Since there were few large centralized states with cities and huge stone monuments in Africa, the continent was usually perceived as "primitive." Descriptions of Africa were used with particular frequency in the European debate about humanity's place in nature and history. The predominant view, especially in the eighteenth century, was that humanity was placed in a divinely given world order, "the great chain of being," which spanned from stones, via plants, animals, and humans, to God. The very idea invited quests for "missing links" between different types of being. In philosophical speculation at the end of the seventeenth century and during the eighteenth century, Africans—especially Hottentots—were singled out as the lowest form of humanity, bordering on chimpanzees and orangutans (Lovejoy, 1936:227 ff.).

With the colonial conquest of Africa at the end of the nineteenth century, a partly different picture of Africa was gradually drawn. Information was communicated to Europe by travelers, administrators, and missionaries, who described Africa in European conceptual terms as the opposite of modern, industrialized Europe. Despite the greater knowledge of Africa available in Europe, the primitive stamp was further confirmed, precisely because Africa "lacked" obvious equivalents to European economy, politics, and justice. Since these categories could not be found in their pure European forms, the Africans remained "primitive" in the evolutionary perspective that was created in the nineteenth century. The continent was thought to have stagnated at an early phase of development, so it lacked history. Consequently, Africa was not studied by historians but by anthropologists, and it was Africa that provided much of the basis for anthropological theory (Holl, 1990; Kuper, 1988; MacGaffey, 1986).

Through anthropological fieldwork, large quantities of artifacts were assembled and brought back to the colonial powers in Europe. In keeping with the view of Africa as primitive, these artifacts were exhibited in ethnographic museums but not in museums of "fine art." In the ethnographic museums, however, the African artifacts were "discovered" around 1900 as an exotic, "timeless" art by European artists. Stimuli from this *art trouvé* played a decisive role in the emergence of modern art, such as Picasso's and Braque's cubism. "Primitive" African art was viewed by these artists as an alternative to the decadence of European art, and the "intuitive expressionism" of the African objects was emphasized (Rhodes, 1994; Rubin, 1984). African myths, which at the same time became available in European translations, could likewise be appreciated in the same way. For Freud, African myths were clear expressions of the universality of humanity (Chrétien, 1986).

Proper archaeological interest in the past of Africa was aroused in the 1880s, at the same time as the final colonization. That was when colonial officials and officers, on journeys and "pacifying expeditions" in the new colonies, came across stone tools of a great age. These amateurs collected finds, did occasional excavations, and sent artifacts back home to ethnographic museums in their own countries (de Maret, 1990; Holl, 1990; Kense, 1990; Robertshaw, 1990b). With few exceptions, attention was focused on the Stone Age, chiefly the Paleolithic. Later periods were not considered interesting, since Africa was regarded as a stagnant continent, which had stopped at an early stage of human development and could only be set in motion again by impulses from outside. Any expressions of "civilization" that were actually found, such as cities, stone monuments, and realistic art, were seen as a result of external influence, whether from Phoenicians, Jews, or Arabs (Figure 19). The Arabs in particular were viewed as the predecessors of the Europeans when it came to "civilizing" Africa (Holl, 1990).

Professional archaeology was established in several African colonies in the 1920s and 1930s, by Europeans with training in archaeology. The racism was toned down, but archaeology followed the tradition of previous generations of amateurs. The sights were focused on the earliest phase of the Stone Age, since this period was seen as Africa's most important contribution to human history. A typical expression of this trend is the work of Mary and Louis Leakey in Olduvai in the interwar years (de Maret, 1990; Holl, 1990; Kense, 1990; Robertshaw, 1990b). Later periods in African history, on the other hand, were rarely studied by archaeologists or historians; they were the province of anthropologists, who developed functionalism partly on the basis of African field studies. Even in the functionalist perspective, however, Africa remained "primitive," since the continent largely lacked functional criteria of civilization, such as writing and cities (Kuper, 1988).

From the 1930s there were nevertheless occasional signs of change in archaeology and in the view of Africa's past. French archaeologists excavated in historically known cities in West Africa, while British archaeologists investigated Great Zimbabwe (de Barros, 1990; Hall, 1990). At the same time, African intellectuals, trained at European universities, began to criticize the European view of Africa's timeless history. For these intellectuals it was important to restore African dignity by studying the history of Africa in a new way (de Maret, 1990; Holl, 1990).

It was not until the end of the 1950s, however, that the view of

Figure 19. A British expedition at Great Zimbabwe in 1891 (Bent, 1892:153). Theodore Bent's expedition to "the ruined cities of Mashonaland" is an early example of the interest in the large stone monuments of Zimbabwe, and also of a European attempt to explain them in terms of foreign influence. On the basis of an unscientific sample of artifacts, Bent believed that the monuments of Zimbabwe were erected as fortified points of support by Arabs who extracted gold in southern Africa for export to Egypt and Phoenicia in the first millennium B.C. (Bent, 1892:220 ff.). "Great Zimbabwe soon became a symbol of the justice of European colonization, which was portrayed as the white race returning to a land that it had formerly ruled" (Trigger, 1989:131).

Africa's past changed radically, both inside and outside the continent. In several old colonial powers, "African history" was established as a special subject, and the oral tradition simultaneously began to be studied as a special African historical tradition (Coquery-Vidrovitch and Jewsiewicki, 1986; Lovejoy, 1986; Moniot, 1986). Archaeology changed direction in earnest with the political liberation of most African colonies around 1960. The new states deliberately sponsored archaeology, even though the work in the following decades continued to be done by Europeans and Americans. Instead of the Paleolithic, the focus was now on the African Iron Age, that is, the history of the last 2,000 years. The express aim was to create a noncolonial African history and identity with the aid of archaeology (de Maret, 1990; Holl, 1990; Kense, 1990;

Robertshaw, 1990b). Besides historiography, an important aspect of this identity construction was the active preservation of historical settings and monuments (see e.g., Ghaidan, 1976). A parallel to this change of archaeological scene is the renewed study of "timeless" African "art." Instead of regarding the objects through a European aesthetic, many studies now sought to reconstruct the historical, functional, and symbolic contexts of the objects (e.g., Fraser and Cole, 1972; Vansina, 1984).

Indigenous African origins have continually been emphasized in all African archaeology since 1960. Trigger (1990) has pointed out that this emphasis has become particularly distinct since African nationalism has been combined with an archaeological tradition, new archaeology, that stresses the significance of internal change. Yet the nationalistic features of African archaeology—and history—are not without problems. As early as the 1970s, there was criticism that Africa's past was mainly studied and assessed using European yardsticks. States, cities, art, and trade were singled out as important aspects of Africa's history, but critics felt that other aspects should be studied (Neale, 1986). In line with this criticism, Bassey W. Andah (1995) calls for analyses of archaeological concepts in order to create new categories that can give a greater understanding of African experiences, both now and in the past.

Moreover, archaeological results seldom give legitimacy to today's African states, since the political borders are the result of the colonial division of the continent. Archaeology in fact shows regional connections that cut across today's national borders, which has led several governments to distrust archaeology (Holl, 1990). The situation has not been made any easier by the fact that archaeology at the end of the 1970s was affected by the same economic crisis that struck the rest of the continent (Shaw, 1989). There are African archaeologists today who would like to tone down the nationalistic tendencies in African archaeology. Augustin Holl from Cameroon calls for a more humanist archaeology concentrating on general human problems. In relation to the recurrent, politically important question of internal or external origin, he would advocate a compromise. Holl (1990) thinks, for example, that the spread of iron technology over Africa must be seen as a combination of various elements that can be of both external and internal origin.

With the reorientation of African archaeology after political liberation, the scene today is dominated by historical archaeology. Yet the conditions for historical archaeology vary in different parts of the continent, since the possibility of relating material culture to text or oral tradition varies greatly.

From the perspective of text, different study areas for historical archaeology in Africa can be demarcated both chronologically and geographically. The border zone between the Middle East and Africa south of the Sahara comprises the regions with the oldest evidence of writing, namely, Nubia, with texts starting from around 1700 B.C. (see Figure 41), and the Aksum empire and its predecessors, in present-day Eritrea and Ethiopia, with texts starting from at least 400 B.C. In both cases, moreover, there are Egyptian, Greek, and Roman reports about the states (Brandt and Fattovich, 1990; Connah, 1987:24 ff., 67 ff.; Munro-Hay, 1991; O'Connor, 1990). Another area comprises the West African savannahs and the East African coast, with kingdoms and cities known chiefly from Arab sources from the ninth century A.D. (Connah, 1987:97 ff., 150 ff.). In connection with the Islamization of these areas, Arabic script came into use, thus establishing the requirements for a locally based historiography. In West Africa, historians from Timbuktu and Jenne are particularly known, and the Kilwa Chronicle from around 1530 represents the oldest historiography in East Africa (Djait, 1981; Hrbek, 1981). A third area consists of regions with no writing before colonialism, but with spectacular traces of the past that attracted early European interest. These include the rainforest belt of West Africa, with Benin and its realistic portraiture, and southern Africa with the huge stone monuments of Zimbabwe (Connah, 1987:121 ff., 183 ff.). A fourth area, finally, is represented by European colonial settlement in Africa, with the fortified trading stations along the coasts and the Boer settlements in the Cape Colony (DeCorse, 1993; Hall, 1993).

Despite great chronological and geographic differences between the objects of study, text-based archaeology has been pursued in a similar way over the continent. Archaeology has mainly been concerned with studies of politically important monuments and sites, especially cities, which are often mentioned in contemporary texts. In all cases the archaeological studies have shown that the cities were much older and more complex than was evident from the written sources. Landscape surveys have emphasized how the cities grew out of the surrounding African hinterland. Some of the archaeological findings have meant that the indigenous origin of urbanization has been stressed at the expense of foreign influence. The distinct Muslim character of many West African and East African cities can thus be seen more as a gradual Islamization than as an original feature (Connah, 1987; McIntosh and McIntosh, 1984; Munro-Hay, 1991; Sinclair, 1987). In view of the fact that early African cities, such as Jenne-jeno in present-day Mali, deviate from the "normal" picture of a city, the archaeological findings have also been seen as important African contributions to a general discussion of urbanism (Fletcher, 1993; McIntosh and McIntosh, 1993). Since archaeology has often been pursued in close collaboration with history, to create a noncolonial African history, material culture has usually been viewed as a kind of complement to written sources. Archaeology has made it possible to go beyond the temporal horizon of the texts, and it has also been possible to study questions that are not as well known from texts, such as production and social organization.

A somewhat different profile is shown by the historical archaeology that focuses on colonial settlement, partly because the contemporary European texts are so extensive and varied. This is especially clear in South Africa, where archaeology studying Boer settlement in the Cape Colony was established in the late 1970s, partly inspired by historical archaeology in the United States (Hall, 1993; Schrire, 1991). The focus of archaeological attention has been on colonial architecture, the slave population in Cape Town, and the influence of the colony on the original inhabitants of the area. The subject is particularly interested in the origins of white repression in South Africa, emphasizing the underclass and all the unknown people with no access to writing.

There are, however, methodological difficulties in bringing out the unknown sides of the Cape Colony. Martin Hall (1993) underlines that the form in which archaeology is presented depends in large measure on whether or not there are contemporary texts. Material culture with no connection to texts is often reduced to bar charts and statistics, whereas objects that can be associated with written sources can be described in more qualitative terms and sometimes can even be related to particular individuals. The narrative form in historical archaeology thus appears to vacillate between the extremes of statistics and biography. It is also difficult to detect in artifacts the underclass and any resistance it put up, since material culture chiefly reflects the dominating colonial upper class. Hall (1994) therefore believes that resistance can only be detected by deliberately looking for differences between artifacts and texts.

Historical archaeology in Africa can also be regarded in another way, however. If the relation between material culture and oral tradition is emphasized (see Schmidt, 1978:3 ff.), instead of the relation between artifact and text, the African situation can be described in a quite different way. As I have pointed out before, oral tradition has been regarded since the end of the 1950s as a special African historical tradition, often held up as a criticism of the text-based Western view of history. The opposition of "prehistory" to "history," and of literate societies to nonliterate societies has been criticized for being an ethnocentric concept, with "pejorative connotations that denigrate the African historical experience" (Schmidt, 1983:64).

The focus on oral tradition has led to detailed studies of its character and of its relation to material culture (Hampaté Bâ, 1981; Vansina, 1965). Yet the significance of oral tradition in African historiography has also meant that historical archaeology can be carried on over the whole continent, regardless of whether texts exist. In practice, oral tradition is often used in combination with texts. Oral tradition has been used in several different archaeological contexts and in studies of widely varying temporal depth; therefore, studies in historical archaeology sometimes verge on ethnoarchaeology.

In connection with archaeological surveys, oral tradition has been important for localizing and identifying sites and determining their functions. Oral tradition also plays a decisive role in studies of cultural practice (Schmidt, 1990). Information from living people has been used to study the utilization and perception of space in the town of Lamu in present-day Kenya, with "Swahili" houses still surviving (Donley, 1982). Oral tradition has likewise been used for cognitive interpretations of the central stone monument in Great Zimbabwe (Huffmann, 1986) and to provide a cosmological background to the settlement pattern in southeast Nigeria (Darling, 1984). Another recurrent archaeological theme that has been studied with the aid of oral tradition is iron production (Figure 20). Several scholars have pointed out the sexual metaphors associated with iron production, with furnaces shaped like female bodies, and with parallels in the decoration of furnaces, women, pots, and clay figures (Collett, 1993; Matenga, 1993).

In the relation between oral tradition and material culture, it is above all the agreements that have been emphasized, with the aim of reconstructing cultural practice. Yet several scholars have criticized the use of oral tradition, especially for the interpretation of precolonial remains, because colonialism radically changed Africa. What might seem to be cultural practice of high age may instead be a result of colonialism, with little relevance for the interpretation of precolonial remains. Because of these problems, Ann Brower Stahl (1993) urges that comparisons between material culture, oral tradition, and ethnographic descriptions should be as close in time as possible.

Alongside studies of cultural practice, a special feature in African historiography is the attempts to write political history with the aid of archaeology, linguistics, and oral tradition (e.g., Oguagha and Okpoko, 1984; Pikirayi, 1993). In the 1960s there was tremendous optimism about the possibility of writing political history with the aid of archaeology, especially after Jan Vansina (1965) presented methods for the use of oral tradition in writing history. Vansina believed that it was possible to evaluate the credibility of oral tradition. Normally it was reliable for the last 200 years, and could sometimes go back as much as 500 years; he claimed that archaeological excavations could occasionally confirm that an element of oral tradition was more than a thousand years old.

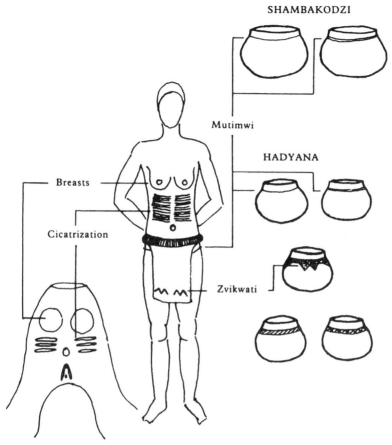

Figure 20. Parallelism in the decoration of iron furnaces, women, and pottery in southeast Zimbabwe (Collett, 1993, fig. 31.2, by courtesy of Routledge). The parallelism, which can sometimes be confirmed by oral tradition, reveals a precolonial cognitive system in which iron production is equated to childbearing and cooking.

Oral tradition seems to be particularly reliable in areas with centralized states. Several early archaeological studies therefore sought to verify and date oral traditions (Schmidt, 1990).

Today the attitude is more restrictive, since the relation between oral tradition and archaeology has proved to be more complicated than previously thought. It is very difficult to obtain archaeological "confirmation" of an oral tradition, since it is so vague and indeterminate. Disparate elements may often have been incorporated in what seems like a coherent tradition. Excavations of "royal" graves in Rwanda are one example. According to oral tradition, they date from the mid-seventeenth century, but archaeology has been able to date them to the third century B.C. The striking discrepancy can be explained by the fact that dynasties have successively adopted and integrated older traditions about graves in their mythologies to legitimate their power. Peter R. Schmidt (1990:270) therefore sums up the attempts of the last 30 years at correlation: "If we have learned anything of importance in this endeavour to marry oral tradition with archaeology, it is the lesson that oral traditions often identify processes of transformation, and are symbolic renderings of historical manipulations that mask changes in political and economic power."

AMERICA

Historical archaeologies in America are numerous and varied. The subjects are primarily concerned with the pre-Columbian states in Central and South America, but the period starting with the European conquest has also been studied archaeologically, not least in North America. Archaeology in America is of European and North American origin. With the increasing economic and political dominance of the United States in the twentieth century, archaeology in the Americas has been heavily influenced by archaeologists from the United States, although there are indigenous archaeological traditions in Central and South America.

The ultimate origin of Western interest in America and its early cultures goes back to the actual "discovery" of America. Ever since Europeans around 1500 began to realize that America was a completely new continent, hitherto unknown to them, America has been a problem in terms of the history of civilization. The question was how the Indians could be incorporated in Christian historiography. Most debaters thought that the inhabitants of America originated in the Old World; the Indians could be seen as descendants of Jews, Phoenicians, or Egyptians (Bernal, 1980:19 ff.). Some people, however, claimed that the Indians were not necessarily of Old World origin; they instead represented humans "outside" the biblical tradition. Giordano Bruno (1548–1600) was therefore able to use the existence of the Indians and information about their infinite perception of time as a radical critique of the Bible and the short biblical chronology (Schnapp, 1993:225 ff.). These different speculations about the true origin of the Indians often served as a background to early archaeological work in America in the nineteenth century. With the breakthrough of professional archaeology in the early twentieth century, however, ideas about intercontinental contacts over the Atlantic were relegated to the margins of scholarly discussion. Links across the Pacific have occasionally been investigated instead (Willey and Sabloff, 1974:165 ff.).

There are interesting historical archaeologies in all the Americas, concerning both the pre-Columbian era and the colonial period (cf. e.g., Funari, 1995; Politis, 1995; Schaedel, 1992; Willey and Sabloff, 1974). I have nevertheless chosen to present only three of these traditions. They are in some ways very different, but they are all geared to studies of societies that dominate or have dominated parts of the continent. They are the "historical" parts of Mexican and Peruvian archaeology and historical archaeology in the United States.

America's Greece: Mexican Archaeology

The preconditions for historical archaeology in Mexico are numerous and varied. When the Spaniards conquered Central America in the 1520s, four different written languages were in use. Writing had been used in the area since at least 500 B.C. and continued in use until 1600, when more far-reaching Hispanicization began. From the sixteenth century there are many important Spanish eyewitness accounts, and later an indigenous Spanish literature arose, preserving elements of the pre-Columbian tradition (Bernal, 1980:35 ff.; Marcus, 1992: 35 ff.).

The Spaniards initially showed a keen interest in the Aztec Empire and its neighbors, but with active Hispanicization from 1600 onwards, Mexico's pre-Columbian history was defined out of existence. Cortez's conquest was viewed as the start of Mexico's history, and to underline this Eurocentric attitude, many of the Aztecs' monuments were deliberately destroyed. By the end of the seventeenth century, however, a special Mexican Creole culture can already be detected, and this became more distinct with the growth of Mexican nationalism in the latter half of the eighteenth century. In this Mexican culture, borne by indigenous intellectuals, who were often mestizos, there was increasing interest in the early states. In their recurrent critique of the Eurocentric worldview, the Mexican intellectuals claimed that the pre-Columbian cultures were a part of Mexican history (Bernal, 1980:49 ff.).

European interest in Mexico's pre-Columbian history was aroused in earnest with the works of Alexander von Humboldt (1769–1859), based on his travels in America in 1799–1804. The interest shown by Europeans and North Americans in the early history of Mexico grew further with the political liberation of the country in 1821. From the

1840s onward, English and American explorers searched the jungles of Yucatán for unknown cities and temples (Figure 21). The most famous were John Lloyd Stephens and Frederick Catherwood, who described their adventurous trips in books that gained widespread popularity (Bernal, 1980:103 ff.; Stuart, 1992).

An archaeological museum was founded in Mexico City in 1826, shortly after independence. It was not until the 1880s, however, that the subject of Central American studies was gradually professionalized, with linguists, historians, and anthropologists from Europe and North America devoting their efforts to the early history of Mexico. From around 1900, the major Mexican monuments began to be studied and restored by Mexican scholars, in a desire to manifest the greatness of Mexico (Bernal, 1980:142 ff.; Vázquez León, 1994). Not until the 1920s, after the Mexican Revolution, however, did excavating archaeology make its breakthrough, with stratigraphic investigations. A characteristic split in archaeological work can be discerned at this early stage. On the one hand, indigenous Mexican archaeologists, such as Mañuel Gambio, studied the central highlands around Mexico City. Their interest was focused on the Aztecs and their predecessors, such as the Toltecs and the city of Teotihuacán, since it was this part of Mexico's history that was most important for creating a national identity. At the same time, museum areas were established round the great "national" monuments, which were often insensitively restored (Bernal, 1980:160 ff.; McGuire, 1992:62 ff.; Vázquez León, 1994).

The Maya region, on the other hand, was totally dominated by American archaeologists, and in the period 1914–1958 dominated by a single excavating institution, the Carnegie Institution in Washington, D.C. Maya culture was not as interesting from the Mexican point of view, since it was on the periphery of the country and partly beyond the borders of Mexico. From the North American point of view, however, the Maya were of great interest, since they were traditionally viewed as the "advanced" and "civilized" mother culture of the other Central American cultures. The interest in the Maya was thus a kind of quest for "America's Greece" or "the Athenians of the New World" (Bernal, 1980:173 ff.; Hammond, 1983; Marcus, 1992:3 ff.; Schávelzon, 1989; Stuart, 1992).

In the interwar years, both North American and Mexican archaeologists were mainly concerned with chronology, inscriptions, and individual monuments, but in the postwar period the American archaeologists have extended their interest to settlement and landscape as a whole. This development came because several American archaeologists working in Mexico were active in developing processual archaeology in the 1960s. In the highlands and lowlands alike, North American

Figure 21. The "House of the Dwarfs" temple in Uxmal in 1882 (Charnay, 1887:403). The French explorer Désiré Charnay was one of many Europeans and Americans in the latter half of the nineteenth century who traveled in Central America in search of vanished cites and kingdoms. Thanks to their pictures and measurements it is possible to gain some idea of what the Central American monuments looked like before the restorations of the twentieth century.

archaeologists since around 1960 have studied the ecological and agrarian bases of the civilizations in landscape surveys lasting for many
years (Flannery, 1982; Sanders et al., 1978). The view of Mesoamerica's
central places has changed too, as large areas of ruins have been excavated and charted since the 1950s. Places with large temples that had
long been famous, such as Tikal in the Maya area, have proved to be big,
multifunctional cities (Sanders and Webster, 1988). Many of the results
of archaeological excavations in cities and rural areas have also been
used in comparisons of the history of civilization. Since the Central
American civilizations are historically separate from the cultures in the
Old World, the Aztec Empire and the Maya area have frequently served
as points of reference in the search for similarities and differences in
the emergence of "complex societies" (cf. Adams, 1966; Trigger, 1993;
Wheatley, 1971).

The oral tradition in the Maya area attracted attention as early as
the 1920s (Hammond, 1983), but in the postwar period early records of
tradition and present-day oral traditions have increasingly been integrated in archaeological work, to create history rather than timeless
anthropology. Oral tradition is often used in studies geared to the
history of an entire area, including the colonial period as well (Ball,
1986; Flannery and Marcus, 1983).

Mexican archaeologists have continued their work in the highlands
around Mexico City, such as their mapping of the metropolis of Teotihuacán and the remarkable excavations of Templo Mayor in Mexico
City (Townsend, 1992:144 ff.; Willey and Sabloff, 1974:165 ff.). Unlike the
North American archaeological tradition, however, this Mexican archaeology has been influenced very little by processual archaeology and
is still associated with a tradition focusing on cultural groups and
descriptive chronologies. According to Luis Vázquez León (1994:85), this
is because it is "the perfect theory for the exhibiting and public function of monumental archaeology."

In methodological terms, the most interesting part of postwar Central American archaeology is Maya studies, since they represent a
historical archaeology in the making. The decipherment of the Maya
script has been in progress since the 1880s, but it has gone much more
slowly than the decipherment of ancient Middle Eastern scripts. It is
only in the last 10–15 years that there has been a breakthrough (Figure
22), when most scholars finally accepted the idea that the Maya hieroglyphs contain phonetic elements (Coe, 1992; Stuart, 1992). Older interpretations can thereby be confronted with completely new written
sources. In this encounter there are several thought-provoking differences in the perception of the Maya culture. In the first half of the

twentieth century the Maya culture was seen as peaceful theocracies governed by priests who resided in ceremonial centers that were otherwise only temporarily inhabited. This picture was already modified by archaeological investigations in the 1950s and 1960s, which showed, among other things, that the temple sites were large cities. The deciphered texts, however, reveal the Maya culture to have consisted of aggressive political units in constant conflict with one another. As among the Aztecs, human sacrifice was an important religious ritual (Culbert, 1991; Fash, 1994; Marcus, 1992). The ecological foundation of the Maya area was stressed in the 1960s and 1970s, but with the texts now being accessible, ideological and political explanations have come to the center of the discussion of Maya culture (Fash, 1994).

There has been extensive debate about the political organization of the Maya. In 1946 Sylanus G. Morley maintained, on the basis of analogies with ancient Greece, that Maya culture represented a number of city-states. The recent decipherment of the script, and especially the studies of the emblems denoting different places and their territories, have largely confirmed Morley's view (Culbert, 1991:1 ff.). The sizes and mutual relations of the city-states, however, have been viewed in very different ways. Joyce Marcus (1976, 1992:153 ff.) sees Maya culture as a few hierarchically organized states, while Peter Mathews (1991) argues that the area corresponds to 60–70 small and equal city-states. The script, the pictorial art, and the architecture were held in common by the city-states, whereas material culture of a simpler character, such as pottery, showed local variation. Archaeological investigations have thus uncovered a complex picture of the Maya city-states, with a shared elite culture and a variety of local cultures representing people outside the elite (Sharer, 1991).

Since 1990 there have been several attempts to write the political history of the Maya, but these have been accompanied by the first critique of the new historiography. Joyce Marcus (1992:3 ff.) claims that the Maya texts should be viewed more as propaganda than as historical sources, which means that they cannot be immediately used to reconstruct the political history of the area. Instead, archaeology can be used to assess the reliability of the texts. For instance, when the ruler "Lord Shield" of Palenque is stated to have lived to the age of 80, although his skeleton has been identified by osteologists as that of a 40-year-old, the discrepancy suggests that he may have been a usurper. He may have destroyed the monuments of his immediate predecessors and replaced them with his own inscriptions, which associated him with older rulers and thus gave him a kind of political legitimacy (Marcus, 1992:303 ff.).

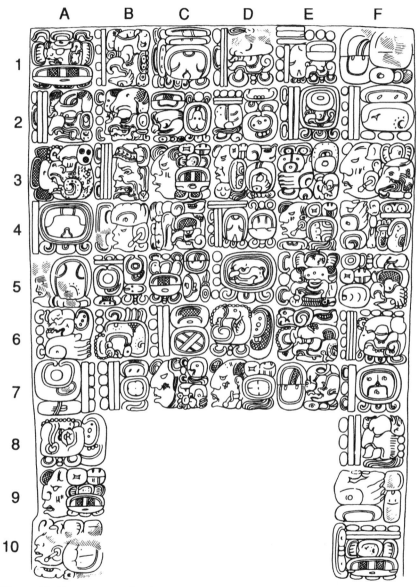

Figure 22. Two versions of stela 3, dated A.D. 711, from Piedras Negras (Marcus, 1992, figs. 10.24 and 10.25, by courtesy of Joyce Marcus). The two figures show how a Maya text (left) may be translated glyph-by-glyph into a modern language (right).

	A	B	C	D	E	F
1	Initial Series Introductory Glyph	9 baktuns [cycles]	0 kins 10 uinals	12 tuns	15 kins 8 uinals 3 tuns	Forward count to
2	12 katuns [units of 20 years]	2 tuns [years]	Forward count to 1 Cib	14 Kankin	11 Imix	14 Yax
3	0 uinals [months]	16 kins [days]	Lady Katun	Lady Akbal [darkness]		Lady Katun [vulture substitute]
4	5 Cib	7th lord of the night	Ruler's title and name	10 kins 11 uinals 1 tun	Lady Akbal [darkness]	Completion of 5th haab
5	—	Moon age is 27 days	1 katun forward count to	4 Cimi	1 katun	25th anniversary of ruler's accession to the throne
6	2 lunations	Glyph X	14 Uo	Was born	Ruler's name	19 kins 4 uinals
7	29-day moon	14 Yaxkin	Lady	Lady Kin [sunlight]	Forward count to	6 Ahau
8	Was born					13 Muan
9	Lady Katun					Completion of
10	Lady Akbal [darkness]					14th katun [A.D. 711]

Figure 22. (*Continued*)

The question of the relation between artifact and text has long been a subject of debate in Mexico. Ignacio Bernal, in a programmatic article (1962), calls for more deliberate analyses of material culture and writing. He believes that texts can solve methodological problems in archaeology, for example, whether there is an association between ethnicity and material culture. Writing could also be a point of departure for archaeological excavations, although archaeology is more suitable for "cultural history." Bernal was really looking for a new type of scholar, the "ethnoarchaeohistorian," since artifact and text "complement each other, thereby producing a wealth of knowledge sometimes unsuspected" (Bernal, 1962:225). Since early colonial sources often play a major role in archaeological interpretation, Diane Z. Chaze (1986) has consciously juxtaposed early Spanish descriptions of the social organization of the Maya area with archaeological investigations of "post-classical" Maya culture (ca. 1300–1520). In this encounter, Chaze can trace differences in the manner of classification. Excavations of graves and settlements suggest gradual social differences in the population, whereas the texts proclaim a more rigid division into three different classes.

As regards the pre-Columbian script in Central America, Joyce Marcus sees it as a political tool used by a small, separate elite who were specially trained to read and write. Like the Egyptian hieroglyphs, the Central American script was used primarily to express political and religious messages. Writing belonged to "noble speech," so we should not expect total agreement with the material culture. Marcus (1992) makes a plea for a more active combination of archaeology and epigraphy in Maya studies, not least to test the reliability of the texts as evidence.

Since the Central American written languages are mainly known through inscriptions on stone, the discussion of the relation between artifact and text has in large measure concerned the physical context of the text (see Figure 22). In several cases there are clear links between new or rebuilt monuments, new inscriptions, and changes in political power (Fash, 1988; Matos Moctezuma, 1988). An example is Tikal, where detailed stratigraphic studies have shown seven large-scale projects to rebuild central parts of the city, marked by new inscriptions and associated with important rulers who are known from inscriptions in other places (Jones, 1991). Inscriptions are also an important element in the design of the big cities, which through the monuments and their placing often symbolized the cosmology and thus expressed the divine origin of the ruler's power (Ashmore, 1992; cf. Wheatley, 1971). An important aspect of the physical context of the texts is their secondary use. In the Maya area many inscribed stelae were buried or used as building material or for other purposes, probably in deliberate attempts to rewrite the official history by making older texts illegible or inaccessible (Marcus, 1992:143 ff.).

The interaction between artifacts, texts, and pictures on individual monuments has also been studied. Janet Catherine Berlo (1983) argues that Western text culture has made us blind to other forms of expression. It is therefore important to try in a new way to see the interplay of architecture, pictures, and texts. The different forms of expression may complement one another, emphasize the spatial configuration, or be a play on words intended to stress the nuances of the language and the picture (see Figure 40). For Berlo, the fundamental aesthetic principle in Central America is to renew an utterance in different forms of expression.

The Past as a Laboratory: Peruvian Archaeology

Historical archaeology in Peru shows both similarities to and differences with Mexican archaeology. On the one hand, there is a similar sequence of monument-building cultures in South America; on the other hand, there was no real written language before the Spanish conquest. Administrative communication in the Inka Empire and its predecessors was based on *khipus*, bundles of strings with knots. By varying the colors of the strings, the position of the knots, and the relative order of strings and knots, the Inka administration was able to use these *khipus* as a notation system. Sometimes the *khipus* were also used together with painted pictures on wooden panels to record historical events. Another significant difference from Central America was that the South American highlands consisted of a single political unit, the Inka Empire, when the Spaniards arrived (Pärssinen, 1992:31 ff.).

There are no written sources, then, until the Spanish conquest in 1532. The texts consist of Spanish eyewitness accounts and early colonial surveys, as well as chronicles and records of the native oral tradition, sometimes written in Indian languages (Figure 23). An important source is Garsilaso de la Vega, who was himself descended from the Inka family and who devoted several works around 1600 to the imperial tradition of the ancient ruling family (Murra and Morris, 1975; Pärssinen, 1992:50 ff.). A special dimension in Peru and the South American highlands is that there has been a strong cultural continuity, despite the long Spanish rule. There are therefore good opportunities for ethnoarchaeological studies (Cordy-Collins, 1983). Generally speaking, however, the state of the sources means that historical archaeology is difficult to pursue beyond the Inka Empire. The *terminus post quem* is

normally the thirteenth century, when the Inka Empire, or Tawantin-suyu ("the kingdom of the four quarters") as it was called, was established as one of several regional states in the Andes. The kingdom did not expand to become an empire until the 1470s (Pärssinen, 1992:71 ff.).

The Spaniards took an interest in the pre-Columbian cultures of South America as soon as they had conquered the Inka Empire. As in Mexico, however, earlier historical periods were defined out of existence during the deliberate Hispanicization in the seventeenth century, and it was not until the liberation from Spain in the 1820s that the pre-Columbian cultures became important for the new national identity. Peru gained political independence in 1821, and only five years later, in 1826, a national museum was established in Lima to exhibit archaeological finds. The greater part of the nineteenth century was nevertheless characterized by unprofessional excavations in search of treasure, the main aim being to build up museum collections both in Peru and abroad (Chávez, 1992).

There was great interest in the remains of the Inka Empire in the nineteenth century, but with the establishment of professional archaeology the focus shifted to the prehistory of Peru. Professional archaeology was introduced in the 1890s by the German Max Uhle (1856–1944), who worked in both Peru and Bolivia. He dug stratigraphically and drew up pottery chronologies, influenced by Flinders Petrie's work in Egypt. Indigenous Peruvian archaeology was established around 1915, with Julio C. Tello (1889–1947) as the figurehead (Chávez, 1992; Daggett, 1992; Morris, 1988). The two pioneers differed greatly in their view of the civilizations in Peru, the disparities being due to their different origins. Uhle used the diffusionist approach in an attempt to trace the Peruvian civilizations to Central America and ultimately to China, whereas Tello, who was closely associated with the Peruvian nationalist movement—*indigenismo*—saw the Peruvian highlands as the origin of the cultures (Matos Mendieta, 1994; Patterson, 1989).

Until the Second World War, Peruvian archaeology was exclusively concerned with the cultures before the Inka Empire, since the Inka Empire was perceived as "history." This archaeology led to the creation of chronological and geographic boundaries for the cultures preceding

←——

Figure 23. A provincial administrator with a *khipu* in his hand, presenting his accounts to the Inka. The two men are surrounded by state storehouses. Illustration in *La Nueva Cronica y Bien gobierno* (1584–1613) by Filipe Guaman Poma de Ayala (after Bjørneboe, 1975:63, by courtesy of Gyldendal). Guaman Poma's great work, with several hundred illustrations, is one example of sources from the early Spanish period that provide important information about the Inka era.

the Inkas (Willey and Sabloff, 1974:116 ff.). A major question since that
time is whether there were older empires in the area. Several archaeologists have pointed to Wari in Peru and Tiwanaku on Lake Titicaca in
Bolivia as conceivable centers of older empires (Schreiber, 1992).

Since the Second World War, archaeology in Peru has been characterized by foreign, especially North American, expeditions. In the 1980s,
foreign projects accounted for 80 percent of all excavations. The American excavations introduced new perspectives to Peruvian archaeology.
The first landscape study was conducted as early as 1946, with a large-
scale survey of the Viru valley, led by Gordon R. Willey (Burger, 1989).
As one of the very first surveys in archaeology, this was one of the
foundations for the later "new archaeology" in the United States (Trigger, 1989:282 ff.). At the end of the 1960s and during the 1970s, processual archaeology was firmly established among American archaeologists in Peru, particularly in the study of the Inka Empire. By virtue
of their presence, Peru functioned virtually as an archaeological laboratory for the study of complex societies (Burger, 1989). Many indigenous
archaeologists, however, have reacted against these perspectives by
developing a Marxist archaeology instead, partly by continuing Gordon
Childe's historical materialism (Lumbreras, 1974; Matos Mendieta,
1994; McGuire, 1992; Patterson, 1989). Since the early 1980s, however,
American archaeological activity has declined sharply as a result of
the state of civil war between the government and the revolutionary
movement Sendero Luminoso (Burger, 1989).

An important change in Peruvian archaeology since the Second
World War is that the Inka Empire has been studied for the first time.
In other words, it is only since then that there has been any historical
archaeology in Peru. Several different aspects of the empire have subsequently been studied (Morris, 1988). Archaeological comparisons with
societies before the Inka Empire have shown that the empire used and
developed earlier technical achievements, such as roads, storehouses,
irrigation systems, terrace cultivation, and *khipus* (Burger, 1989).
Thanks to archaeology, the monolithic picture of the Inka Empire in the
historical chronicles has been modified. Instead, we see an empire with
great regional variations and with unevenly spread manifestations of
political power. The road network, which was a symbol of the power and
authority of the empire, crisscrossed the whole empire, with a total of
some 40,000 kilometers of road. Other symbolic political manifestations, such as the famous cyclopean walls and a certain form of standardized pottery, were however confined to the administrative central
places (Hyslop, 1984; Morris, 1988; Morris and Thompson, 1985).

A problem that has received attention since the 1960s is the spatial
organization of settlement and landscape in the Inka Empire. By com-

bining ruins and early Spanish descriptions, several archaeologists have looked for non-European organization principles. Above all, the cosmological aspects of spatial organization have been stressed. The landscape around the capital, Cuzco, has been seen as ritually planned, to serve as a stage for enacting politically important rites and legends (Farrington, 1992; Hyslop, 1990; Morris, 1988).

A special question concerns the economic organization of the Inka Empire, which was initially debated primarily in economic anthropology. Louis Baudin claimed in the 1920s that the Inka Empire was a socialist state, since it lacked markets, with goods instead circulating via large state-owned storehouses (see Figure 23). Some of these ideas were adopted by Karl Polanyi, who saw the Inka Empire as the ideal type of a redistributive society. With archaeological fieldwork in Peru, American archaeologists have been taking part in this debate about complex societies since the end of the 1950s. Through surveys and detailed studies of the empire's storehouse system, they have been able to confirm the state redistribution system in part (Figure 24), but they have also found traces suggesting that the Inka elite deliberately repressed earlier regional markets in the conquered provinces (Earle and D'Altroy, 1989; Le Vine, 1992; Morris, 1992; Morris and Thompson, 1985).

Despite extensive archaeological work in the last few decades, Craig Morris (1988) maintains that most data on the Inka Empire come from written sources.Archaeology has as yet mostly confirmed, expanded, and modified the written testimony. For example, comparisons between excavated storehouses and early Spanish lists of commodities in the storehouses show that only certain categories of stored items, such as pottery and grain, can be corroborated by archaeology (see D'Altroy and Hastorf, 1992). Discussing the principle of artifact versus text, Patricia J. Netherly (1988) stresses the great difficulty of determining what are similarities and what are differences between written sources and material culture. All interpretations and selections of the different kinds of source material are based on traditions and postulates that often vary from subject to subject. Comparisons between artifact and text are thus not primarily of different sources but rather of different traditions of abstraction associated with different disciplines.

The New Europe: Historical Archaeology in the United States

In North America, as in South America, the very precondition for text-based historical archaeology lies in the conquest and colonization of the continent. Unlike both Central and South America, however, the

N

Europeans did not encounter any large monument-building states in North America, which meant that the North American Indians were regarded in the same primitive and static way as the Africans. When systematic study of the Indians and their history began from the middle of the nineteenth century, it was consequently pursued as a branch of anthropology, not of history. Professional archaeology was established at the end of the nineteenth century, almost as an extension of anthropology. This link between anthropology and archaeology in North America, as Trigger (e.g., 1989:119 ff.) has pointed out several times, can be explained by the fact that the archaeologists were not studying their own cultural past but the past of a "foreign" aboriginal population.

The discipline of history, in contrast, was reserved for the history of civilization in the Old World and for the history of the New World beginning with colonial settlement. In the United States a special "American history," focusing on topics such as the constitution and the founding fathers, acquired a central place in school tuition, as a way to create an American identity. For a long time American history was perceived as white, European history, and it was to this white American history that historical archaeology was connected in its initial stages. It was not until the interwar years that the conditions were created for a historical archaeology geared to the history of North America starting with the first colonization around 1500.

In connection with the depression of the 1930s, the material traces of the United States's colonial past became the object of serious interest for the first time. Jamestown, which was the first permanent English settlement in 1607, became a "Colonial National Monument" in 1930, and five years later old sites and buildings "of national significance" were protected by general legislation (Orser and Fagan, 1995:25 ff.). At the same time, archaeological excavations began at these sites, which were often associated with famous people in the history of the United States (Figure 25). Excavations started in Jamestown in 1934, and thanks to the relief works in Roosevelt's New Deal program, many historical sites and monuments were restored and sometimes reconstructed. This meant that archaeologists for the first time excavated

←——

Figure 24. Plan of the provincial capital of Huánuco Pampa (Morris and Thompson, 1985, fig. 5, by courtesy of Thames and Hudson). Huánuco Pampa, the best-preserved of all Inka cities, was first investigated by American archaeologists in the mid-1960s. The city contains about 4,000 buildings, divided into twelve sectors around a centrally situated rectangular square with a ritual platform in the middle. The city plan shows similarities to the structure of holy places (*huacas*) around the capital, Cuzco (Morris and Thompson, 1985:72 ff.). The long rows of small rectangular and round houses in the southwest part of the city are the remains of the state storehouses. The plan in this part of the city can be described as an archaeological counterpart to Figure 23.

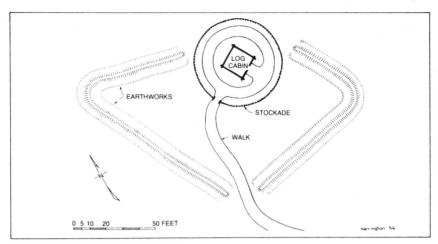

Figure 25. Plan of the reconstructed Fort Necessity, Pennsylvania (Schuyler, 1978:105, after Harrington, 1957, fig. 8, by courtesy of Baywood). In 1754 Fort Necessity was the scene of the first battle between England and France for the control of North America. The British troops were led by George Washington, who became the first president of the United States 22 years later. The fort has been investigated and reconstructed on two occasions, the first time in 1931–1932 and the second time in 1952–1954. As one of the first examples of an archaeologically investigated monument from the colonial history of the United States, Fort Necessity is a good illustration of historical archaeology in the interwar years.

remains from the modern period in the United States. These excavations also gave opportunities to date the European manufactured goods that were usually found at late Indian sites. There had been a lack of good chronologies for many of the typical clue artifacts, such as clay pipes, glass bottles, buttons, knives, and spoons, since archaeologists in Europe had not devoted any attention to this modern material culture (Deagan, 1982; Orser and Fagan, 1995:25 ff.).

Historical archaeology thus emerged as a practical activity in the interwar years. It was not until the 1950s and 1960s, however, that historical archaeology was seriously established and professionalized. It was then so closely associated with the reconstruction of historical sites that several archaeologists preferred the term "historical site archaeology." With the aid of archaeology it was possible to reconstruct important sites and monuments in U.S. history with close attention to detail; these sites included forts, border posts, missions, and early towns. One of the most spectacular projects, which lasted for decades, was the total restoration and reconstruction of the town of Williamsburg in Virginia. The archaeological interest was concentrated throughout on the colonial history of the United States, chiefly its white, European history (Deagan, 1982; Noël Hume, 1969; Orser and Fagan, 1995:23 ff.; Schuyler, 1978).

From the end of the 1960s, however, the dominating white American identity has been increasingly questioned. Many Americans have been "ethnified" by actively searching for, say, an African American or a native American identity in the modern United States (Friedman, 1994). With this "general revival of ethnic pride" (Orser and Fagan, 1995:38), demands for other kinds of historiography have increased. At the same time, there are powerful forces that want to abolish standard courses about Western civilization in schools and universities and replace them with courses about non-Western societies (Friedman, 1994).

Partly as a consequence of the questioning of the white, especially Anglo-American, identity, the character of historical archaeology has changed. The older restoration projects have been criticized for painting an idyllic, conflict-free picture of the American past (Leone, 1973), and at the same time the archaeological interest has been significantly expanded (see Little, 1994). Historical archaeology still has a major focus on the Anglo-American and colonial past of the United States (see Figure 37) (e.g., Beaudry, 1989; Deetz, 1977; Leone, 1984; Shackel, 1993; Yentsch, 1994), but the broader perspective means that historical archaeologists today are also working with such topics as the effect of colonialism on the original population (Lewis, 1984), Chinatowns (Schuyler, 1980), Spanish settlement (Deagan, 1983), slave plantations (Singleton, 1985), black tenants after the Civil War (Orser, 1988a), and more general issues of race, ethnicity, and class (see Little, 1994). Moreover, more recent periods have been studied, such as mining communities (Hardesty, 1988), industrial towns (Dickens, 1982; McGuire, 1991; Wurst, 1991), the consumption patterns of industrialized society (Spencer-Wood, 1987), and waste from modern cities (Figure 26) (Rathje and Murphy, 1992). In the last few years, the gender perspective in various settings has also been emphasized, and the discipline has simultaneously been subject to feminist critique (cf. Seifert, 1991; Spencer-Wood, 1994).

Ever since historical archaeology was established, the character and direction of the subject have been debated. Although several pioneers of the discipline had an anthropological background, they saw themselves as firmly rooted in history. For them the archaeological results were primarily a complement to written sources and a basis for staging the past by means of reconstructions (Harrington, 1955; Noël Hume, 1969). With the broader interest in the U.S. past, the complemen-

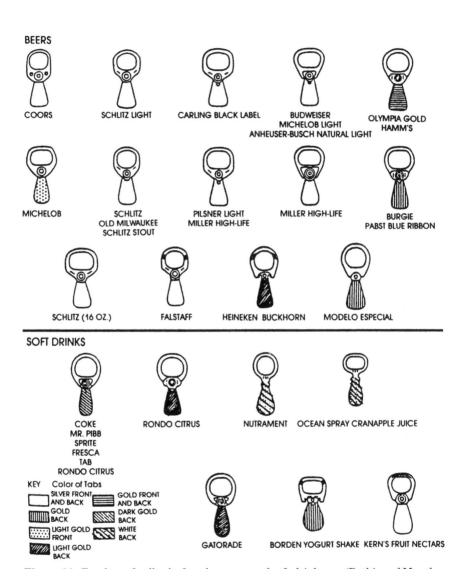

Figure 26. Typology of pull-tabs from beer cans and soft-drink cans (Rathje and Murphy, 1992, fig. 1-D, by courtesy of William Rathje). This typology, which is based on archaeological analysis combined with information provided by can producers, has been used for detailed dating of different layers in modern garbage dumps in the United States. The figure is an example of an extremely modern object of study in historical archaeology.

tary role has likewise been central. Several scholars have expressly declared that the role of historical archaeology is to give insight into little-known aspects of everyday life and to write the history of the invisible people. Although these people did not leave any texts, it is important to study them because they took active part in making history (e.g., Deagan, 1982; Orser, 1988a:246 ff.). In some cases this complementary idea has also been associated with a methodological approach by which the significance of texts in archaeology is generally perceived as subordinate. This perspective was stressed with particular clarity by Stanley South (1977) when he introduced "new archaeology" into historical archaeology.

Alongside the complementary idea, however, there is a partly different tradition, according to which the active meaning of artifacts is emphasized in different ways. Instead of looking for more or less "text-free" zones, scholars have used material culture, for example, to write the historical ethnography of the New England states (Deetz, 1977), to study expressions of power and ideology (Leone, 1984, 1988), and to trace silent everyday resistance among slaves (Singleton, 1985) and industrial workers (McGuire and Paynter, 1991). In these cases, it is emphasized that studies of material culture can give new perspectives on the past, no matter how numerous and ample the written sources are. This emphasis on the active role of artifacts partly anticipates British contextual archaeology, but in recent years it has been further accentuated in a dialogue with the British tradition (Little, 1994). Finally, the role of historical archaeology as an archaeological test area has been repeatedly asserted (see Figure 34). However, even though there are several examples of the way archaeological models and methods have been tested in "controlled" forms, with the aid of extensive written documentation, this methodological profile does not play a prominent role in the subject (Deetz and Dethlefsen, 1965, 1967; Dethlefsen and Deetz, 1966; McGuire and Paynter, 1991; cf. Little, 1994).

Just as there are different views of the role of archaeology, there are also differing opinions about the orientation and demarcation of the subject. Most historical archaeologists work with the past of the United States, but there are also several who want to expand the perspective. Some call for a global historical ethnography for the modern world since around 1500 (Deetz, 1991; Schuyler, 1988). Others have underlined more distinct social perspectives by regarding historical archaeology as a global subject focusing on the capitalist world since 1500 (Leone, 1988; Paynter, 1988). Charles Orser (1996) in particular adopts this stance by doing fieldwork outside the United States. Yet the subject has also been perceived as a "historical material anthropology," valid for the whole world in all periods when writing has existed (Little, 1992, 1994).

Despite the often deliberate championing of the importance of historical archaeology, there is still a paradoxical lack of syntheses to show how fruitful the archaeological perspective is. Historical archaeology has received attention within archaeology, but rarely outside it (see Schuyler, 1988). Mary C. Beaudry (1989) claims that it is because of the "tautological nature" of the subject that historians and anthropologists have hitherto shown little interest in the results. She calls for a much more active archaeological approach to written sources.

The relation between material culture and writing has chiefly been treated on the basis of concrete examples. Several scholars have emphasized, for instance, that it is possible to use waste patterns and artifact categories to create archaeological counterparts to concepts mentioned in written sources, such as ethnic groups and classes (e.g., Deagan, 1983:263 ff.; South, 1977:31 ff.). Others, however, have pointed out important differences between material culture and written sources, since they often reflect different aspects of the past. For example, the stock of artifacts in a household does not look the same if it is defined on the basis of a probate inventory as when it is described on the basis of a refuse heap (e.g., Brown, 1989; Stone, 1989). As a basic principle, Robert L. Schuyler (1978) asserts a fundamental difference between material culture and writing. He claims that the "behavior" of artifacts is directly accessible, in contrast to texts, whose behavior is only indirectly accessible and whose concepts are directly accessible. Another example is Kathleen Deagan (1982:153), who argues that a present-day analysis of material culture differs from an earlier view of the same artifacts, and these views must always be kept apart, just as anthropologists distinguish between anthropology's "etic" classifications and the studied people's own "emic" classifications. Using the analogy of the anthropological concepts, Deagan thus distinguishes between the "etic statements" of archaeology and the "emic statements" of texts from the past (see also Yentsch, 1989).

Important criticism of the programmatic optimism of historical archaeology has also been voiced by Mark Leone, Constance A. Crosby, and Parker B. Potter, who believe that the actual relation between artifact and text has not been problematized to a great enough extent. They stress that there is a fundamental difference between material culture and writing, since the traces are created on different occasions, for different purposes, and normally by different people. Artifacts and texts can thus be seen as different cultures that can be compared with each other. The important thing is therefore to look for differences, or "ambiguities," between artifact and text, which may mean, among other things, that archaeology contributes to "new readings" of texts (Leone and Crosby, 1987; Leone and Potter, 1988).

With this brief survey of historical archaeology in the United States, I round off my global outline of the last three chapters. I thus ignore Oceania, although there are historical archaeology there too, focusing both on the aboriginal populations (Kirch, 1986) and on colonial settlement (Connah, 1988). Some aspects of these traditions will nevertheless be touched upon in the thematic sections that follow. My brief global survey is not complete, whether as regards the world as a whole or the individual disciplines. I nevertheless hope that the sketch has conveyed some impressions from all over the world, showing similarities and differences between the subjects. The following chapter is about these similarities and differences.

The Field of Historical Archaeology

5

INTRODUCTION

The immediate impression of the field of historical archaeology is that of a vast jumble of strictly specialized archaeologies. The disciplines have different histories, the objects of study vary widely in time and place, and the debates are divided. At the same time, however, this jumble is pervaded by a shared feeling among many historical archaeologists that they are working in an intermediate zone, between subjects that are either more object centered or text centered. This feeling of being in-between and having a vague identity is often reinforced by the opinions and demands expressed by other disciplines. Historians have often come with "want lists" of all the things they think that historical archaeologies should be doing (e.g., Finley, 1971; Sawyer, 1983), and many prehistoric archaeologists see the historical archaeologies primarily as a laboratory in which they can test their methods (e.g., Binford, 1977; Clarke, 1971). Moreover, the pioneers of the different historical archaeologies have in many cases been schooled in other subjects, which means that they have perceived the field as having special links with their own original discipline. For example, Michel de Bouard (1969:62), originally a historian, asserts that "medieval archaeology, in particular, must remain an auxiliary science to history." The same view was expressed by the first generation of historical archaeologists in the United States (e.g., Harrington, 1955).

Depending on the scientific context, this intermediate zone has been described in different ways. Alexander Conze in 1869 defined his field as lying between art history and philology, and Hans Hildebrand in 1882 saw himself as working between archaeology and history. Hildebrand's stance is still the dominating one today, especially in Europe. In the United States, the subject is often placed between anthropology and history (Deagan, 1982; Little, 1994). In the same way, the form of scholarly presentation can be seen as oscillating between different genres, depending on the presence or absence of written sources. The historical archaeological text can thus be perceived as lying in a position between "narrative" and "analysis" (see White, 1987:26 ff.) or between "biography" and "statistics" (Hall, 1993).

This sense of "in-betweenness" and dependence on other subjects has also been expressed in various characteristic metaphors. For example, there have been protests against the treatment of historical archaeology as "the handmaiden to history" (Deagan, 1982:158; Hildebrand, 1882:5) or "a poor cousin ... to prehistory" (Little, 1994:29). It has been felt to be difficult to "marry" oral tradition and archaeology (Schmidt, 1990:270), and historical archaeology has simultaneously been regarded both as "the illegitimate offspring of history and archaeology" (Hodges, 1983:24) and as "an orphan branch" between "sister disciplines" (Rautman, 1990:142). The self-definition of the subject has been described as "complex" (Deagan, 1982:153) and the split identity has even been diagnosed as scientific "schizophrenia" (Christophersen, 1992:74).

But what does this sense of in-betweenness involve? Can descriptions of it get beyond these metaphors, which almost evoke associations with a family drama? A crucial starting point for continued discussion is that the in-betweenness is historically conditioned, as an expression of modern human science ever since the middle of the nineteenth century. The specialization of modern science has mainly had an empirical basis, since different types of source material rather than questions and problems have demarcated different disciplines (see e.g., Liedman, 1978). Specialization has resulted in object-centered disciplines such as archaeology, art history, and architectural history, or else in text-centered disciplines such as history and philology. The historical archaeologies, which work with both artifacts and texts, have thus fallen between two stools, and the characteristic feeling of in-betweenness has arisen.

I think that the field of historical archaeology can best be summed up as this dilemma of in-betweenness. On the one hand, the dilemma means that the historical archaeologies are constantly defined in relation to other areas that are either more object centered or more text centered. The purpose of the historical archaeologies is often defined in relation to neighboring disciplines, and through these definitions it is possible to detect attitudes that are not unique to any particular historical archaeology. Instead we see a number of recurrent traditions that are found crossing disciplinary boundaries. On the other hand, the dilemma is linked to the specialization of modern science. With the dilemma of in-betweenness as a point of departure, it is therefore important both to trace the transgressing traditions in the field of historical archaeology and to suggest the broader historiographical contexts in which the field as a whole can be placed.

THE FIELD AS TRANSGRESSING TRADITIONS

A striking aspect of the global survey of the previous three chapters is that, despite the obvious disciplinary division, there are certain recurrent attitudes and approaches in the historical archaeologies. These may be seen in widely differing periods, but they can also occur in several different historical archaeologies at the same time. The thirst for new texts was a major incentive to classical and Mesopotamian archaeology in the 1870s, but the same motive force can still be detected in today's Japanese and Mexican archaeology. In a similar way, the special role of archaeology for studies of, say, economy, has been stressed in almost all the historical archaeologies in the last 30 or 40 years. Although the historical archaeological field is a part of modern science, these attitudes can in some cases be detected long before modern science. The recurrent perspectives all circle around the basic question of why it is important to study material culture even when there are texts from the same periods of the past.

By looking for different attitudes to this fundamental problem, I maintain that it is possible to detect five more or less distinct methodological traditions in the field of historical archaeology: those of aesthetics, philology, history, cultural history, and archaeology. These currents can be traced across the individual disciplines, and on this basis it is possible to describe the field of historical archaeology in a new and different way. Instead of emphasizing the individual subjects, one can demonstrate thematic diversity running across disciplinary boundaries. The different traditions can be followed over long time spans, partly preceding the professionalization of historical archaeology, yet each tradition nevertheless has its own chronological center of gravity. The boundaries between them may be diffuse, but the varying methodological approaches to artifact and text simultaneously show that each tradition has its characteristic features.

Staging the Past: The Aesthetic Tradition

Several historical archaeologies are linked to an aesthetic tradition. This approach is concerned with practical applications of traces of the past, such as the development of historicizing artistic styles, restoration, and reconstruction. As is obvious from the survey of the different disciplines, this materialization of the past was one of the pre-conditions for disciplines such as classical archaeology, medieval ar-chaeology, Japanese archaeology, Mexican archaeology, and historical archaeology in the United States. Moreover, the tradition can also be detected in Egyptology, Indian archaeology, and Chinese archaeology.

Historicizing styles, or deliberate archaizing, in architecture and art are known, for example, from pharaonic Egypt, ancient Greece, and the Aztec Empire. However, the first thoroughgoing historical style was elaborated in northern Italy in the fifteenth century, in what we today call the Renaissance. All historicizing styles require direct study of buildings and works of art that serve as models (Figure 27). This connection between monument studies and the development of a new style was formulated around 1450 by the humanist and architect Leon Battista Alberti (1404–1472) in his treatise *De re ædificatoria* (On the Art of Building). He declares expressly that it is insufficient to read ancient works on architecture, of which Vitruvius's *De architectura* is the prime example. One must study and measure the monuments one-self if one is to create something new: "Therefore I never stopped ex-ploring, considering, and measuring everything, and comparing the information through line drawings, until I had grasped and understood fully what each [building] had to contribute in terms of ingenuity and skill" (Alberti, 1988:154 f.). For Alberti, then, it was clear that the arti-facts took precedence over the texts when it came to practical applica-tions of the past, in the form of classically inspired architecture.

The truly decisive styles for the emergence of classical archaeology, and to some extent of medieval archaeology, were nevertheless the his-toricizing styles that dominated Europe and North America in the late eighteenth century and during the nineteenth century. Since the prin-ciple formulated by Alberti was still valid, the development of the his-toricizing styles required increasingly detailed knowledge of the models (see Figure 2). It was in the quest for an absolute aesthetic for artists that Winckelmann established the foundations for a history of ancient art, and it was in his pursuit of a perfect neo-Gothic architecture that Rickman prepared the ground for studies of English medieval architec-ture. The ideal of historicizing styles thus furthered the emergence of the historical archaeologies.

The same link between historicizing styles and historical archae-ologies can also be detected outside Europe. Napoleon's campaigns in Egypt, which marked the beginning of "orientalist" knowledge of Egypt (see Figure 7), also found aesthetic expression in the French Empire style. Egyptian architecture later served as a model for a neo-

Figure 27. Section of the Roman amphitheatre in Verona, with architectural details from the building, according to *L'architettura* (1537–1551) by Sebastiano Serlio (1457–1554), here from the German translation (Serlio, 1609(3):36). The picture is an example of how an archaizing style such as the Renaissance presupposes a knowledge of material culture in the past, in this case Roman architecture.

Egyptian style, associated above all with solidity, which is why it was used for bank buildings in Europe and the United States (Curl, 1982). In a comparable way, British architects in India created an Indo-Saracen architecture in the latter half of the nineteenth century, through studies of the Islamic art of building in India. A faint echo of the same ideas can also be detected in China during the Cultural Revolution in the 1970s, when the terracotta soldiers in front of the grave of Emperor Qin were held up as ideal models for the depiction of people in the art of socialist realism (see Figure 16).

The ideal of historicizing style has thus led everywhere to increased knowledge of art history and architectural history. When the ideal was abandoned around 1900, with the breakthrough of modernism, aesthetic history was preserved in sciences such as art history and architectural history, partly as a stage in the training of artists and architects. The aesthetic tradition has thus partly detached itself from the original practical applications. Even today, however, it is still possible to see an active union of aesthetic theory and practice, for example, in the relation between advertising and pictorial semiotics (see Sonesson, 1992:67 ff.).

Another aspect of the aesthetic tradition is the care of historic monuments, in the form of restoration and reconstruction. The boundaries between these activities are often fluid. Deliberate restorations are known, for example, from the Roman Empire, and they were important features of Chinese and Japanese architecture. In Europe, the conservation of historical monuments is intimately associated with the ideal of historical styles. For example Alberti (1988:62), as early as 1450, advocated the restoration of ancient monuments rather than demolition. The first truly systematic management of historic monuments came after the French Revolution and the Napoleonic Wars. That was when the exemplary models for the historical styles, such as Greek temples and medieval cathedrals, were restored and reconstructed in earnest (Karsten, 1987:21 ff.).

Because of the direct link with the ideal of historical styles, restoration aimed for purity of style, which meant that the oldest or the predominant style of a monument was highlighted and later additions and changes were removed. The result of this stylistic purity was that the material remains were considered more important than the texts. Restoration required a thorough familiarity with materials, techniques, and styles, as well as a good knowledge of the individual monument (see Figure 5). An illuminating example of the significance of material culture is the rebuilding of Notre Dame in Paris in the 1840s. J. B. A. Lasses and E. E. Viollet-le-Duc write expressly that it was "with the support of reliable texts, drawings, engravings, and above all by deriving information from the monument itself, that we have undertaken this restoration" (cited from Kåring, 1992:163). Thus, restoration and the more profound knowledge of historic monuments further helped to establish the basis for subjects such as classical archaeology and medieval archaeology.

When the ideal of historical styles was rejected around 1900, the principle of stylistic purity in restoration was abandoned. Since then, the emphasis has been on the entire architectural history of the monuments, including the settings in which they stand. The management of monuments has thus preserved a link with the historical archaeologies. The practical work of restoration has been important for the development of methods in a subject such as medieval archaeology (Figure 28). It is by means of extensive and detailed restorations that the stratigraphic principles for standing buildings have been gradually developed (e.g., Francovich and Parenti, 1988; Rodwell, 1981). The conservation of monuments is also the origin of a historical archaeology that I did not cover in the survey above, namely, industrial archaeology. One of the incentives for the discipline in Britain was a controversy about the demolition of Euston Station in London in 1962. Industrial museums, such as the Ironbridge Museum in England, are also typical of the orientation of the subject. In several countries the preservation of the industrial heritage is in fact the most important part of the discipline (Harnow, 1992).

Outside Europe we find a corresponding link between the conservation of monuments and the historical archaeologies. Indian archaeology is partly based on restoration work and surveys of the country's Hindu, Buddhist, and Muslim monuments, which were started in the last decades of the nineteenth century. In a similar way, part of the origin of Japanese historical archaeology may be sought in restorations and studies of architectural history starting in the 1890s. That was when the study of Buddhist temples began in the old imperial city of Nara (see Figure 18). In Mexico, indigenous archaeology is very closely tied to the 160 or so "archaeological zones" established since 1905 (see Figure 21). In these open-air museums the country's past is presented, often in excessively restored pre-Columbian monuments, especially temples (Vázquez León, 1994). Finally, the roots of historical archaeology in the United States may be sought in the restoration and reconstruction of colonial settings (see Figure 25). From the 1930s onward, archaeology was used as a basis for rebuilding and restoration projects, with Colonial Williamsburg as the most magnificent example. As we have seen, the link between archaeology and reconstruction was so close that several archaeologists in the 1960s still preferred the term "historical site archaeology" as the name of the discipline. In more recent years there has been criticism of the idealized reconstructions of the U.S. past

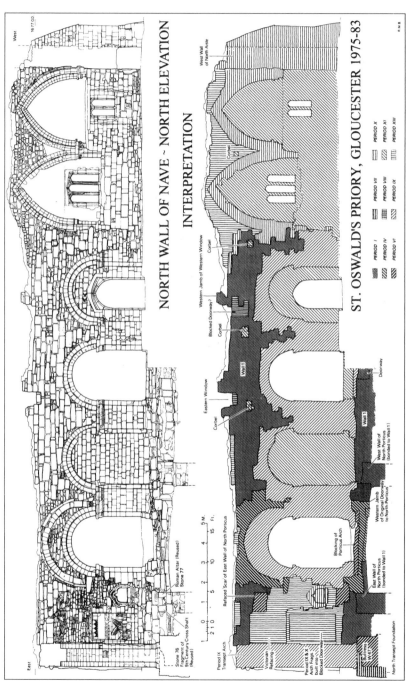

Figure 28. Measurement and stratigraphic analysis of the north wall of St Oswald's Church in Gloucester (Heighway and Bryant, in press, by courtesy of Carolyn Heighway and Richard M. Bryant). By means of the detailed stratigraphic analysis of the church (top) it is possible to distinguish masonry ranging from the start of the tenth century until the nineteenth century (bottom). The investigation exemplifies how architecture in recent decades has been analyzed with archaeological methods.

(Leone, 1973), and this criticism has direct parallels in the many critical views aired about the ideal of stylistic purity in restorations in Europe in the nineteenth century (Kåring, 1992).

With the rejection of stylistic purity around 1900, there was also a change in the view and interpretation of the historical monuments. When they lost their immediate meaning as aesthetic models for historicizing styles, it became interesting instead to study their meaning in the past. In Europe, as in India, the iconography of architecture has attracted attention in earnest since the interwar years (see Figure 14). In regions such as Egypt, Mesopotamia, China, Mexico, and South America (see Figure 24) too, the meaning of city plans and monuments has been studied for a long time (cf. Wheatley, 1971). This long tradition of studies of the meaning of material culture anticipated contextual archaeology with its stress on the meaning of artifacts, but despite this it has received little attention in general histories of archaeology.

The aesthetic tradition in studies of material culture from literate periods can be followed from the Renaissance onward, but it was in the nineteenth century and the first half of the twentieth century that it played a decisive role in the establishment of several historical archaeologies. The explicitly historicizing styles have long since disappeared, and the occasional "historical quotations" in postmodern architecture have no methodological significance for archaeology (see Pedersen, 1990). However, restorations and new ways of staging history, such as museum reconstructions, are still of some methodological significance for the historical archaeologies. In addition, the practical application of the past will probably acquire renewed significance in the future. I am thinking here of multimedia technology and the possibility of visiting a vanished epoch by means of virtual reality.

To sum up the aesthetic tradition, I would first of all emphasize its primarily material profile. Historical styles as well as restorations and reconstructions are ways of staging the past, and material culture has automatically taken precedence over texts here. Since the past is given practical application, there has been no need for any translation into text; the expressions have remained in a material discourse. It is not until a practical application has been transformed into an aesthetic history that translations into text have been necessary. Yet even in these versions, material culture has been taken for granted as the starting point for the textual renderings.

Giving Meaning to Writing: The Philological Tradition

The philological tradition refers to the study of material culture in a philological context. Archaeology serves here as a producer of texts, a source of background knowledge, and a reference point in historical linguistics. Since a large proportion of the texts from the first 4,000 years of our more than 5,000-year history of writing come from archaeological contexts, the philological tradition has played a major role in the field of historical archaeology. This is one of the preconditions for archaeological disciplines working with the Mediterranean, the Middle East, India, and China. Elements of the tradition can also be detected in medieval archaeology, Japanese archaeology, and Mexican archaeology.

Early inscriptions are known to have been collected during the Sung dynasty in China (960–1279). In Europe, however, large-scale collection of Roman and Greek inscriptions did not begin until the Renaissance in the fifteenth century. The most famous of the inscription collectors was Ciriaco de Pizzicolli (Cyriac of Ancona, 1391–1452), who spent 25 years traveling in Greece, copying and collecting classical inscriptions (Stoneman, 1987:22 ff.). Postclassical Christian inscriptions did not attract any attention until the sixteenth century (Weiss, 1969:145 ff.). This epigraphic interest in ancient inscriptions spread over Europe and was later turned toward other than classical languages (Schnapp, 1993:121 ff.). For example, in the seventeenth century there was extensive epigraphic documentation of runic inscriptions in Sweden (Figure 29) and Denmark (Klindt-Jensen, 1975:14 ff.; Randsborg, 1994). Non-European inscriptions were later documented too, even if the writing could not be deciphered; examples are cuneiform inscriptions, hieroglyphs, and the Maya script.

It was not until the first half of the nineteenth century that surveys and excavations began to yield early texts in any numbers. This was when large masses of texts in extinct languages were dug up and collected, especially hieroglyphs, cuneiform tablets, and early Indian inscriptions. From the 1830s on, these texts began to be deciphered, thus becoming accessible for philological and historical study. The role of archaeology as a producer of texts has continued even since this introductory phase. Surveys and excavations have uncovered new texts in both known and formerly unknown scripts, such as Maya hieroglyphs,

---→

Figure 29. Runic stone from Ala in Vassunda, central Sweden, according to *Monumenta Sveo-Gothica* (1624) by Johannes Bureus (1568–1652), here from a posthumous reprint (Bureus, 1664, fig. 9). Inspired by epigraphic studies of classical languages, Bureus began in the 1590s an extensive inventory of runic monuments in Sweden. His inventory is an early example of the way epigraphy in Europe began to be applied to languages and scripts other than Greek and Latin. Bureus's inventory of runic stones was intended to glorify the past of his native land, and also to further runic script as an indigenous writing system, since runes were still used at that time by peasants in remote parts of Sweden.

Linear B, Chinese oracle bones, Japanese *mokkan*, and Russian birch-bark letters.

The important thing in this context is that the hope of finding new texts has often been an important driving force for new excavations. A part of historical archaeology has thus grown up in more or less direct association with the quest for new texts. Cuneiform texts from excavations in Mesopotamia were long perceived as "by far the most important result" (Lloyd, 1980:164). In classical archaeology too, with its long epigraphic tradition, the hope of finding new texts was long a powerful incentive to excavation. In 1875, after the first digging season at Olympia, Ernst Curtius (1814–1896) was still able to write that "it is the written word which speaks most clearly" (cited from Stoneman, 1987: 263). A similar attitude can be detected in China, where large-scale excavations were conducted in Yinxu, since this was the place with the oracle bones, the oldest evidence of Chinese writing. Today, when the Maya hieroglyphs are being deciphered (see Figure 22), we still hear calls for excavations deliberately geared to finding Maya texts, with the purpose of creating history (Culbert, 1991:311 ff.).

Since the large masses of text began to be collected and deciphered in connection with the establishment of philology as a modern science, the texts were studied on the basis of purely linguistic criteria. Although the texts were the fruit of archaeological work, they were studied and classified into different genres with no regard to the find context. It is therefore difficult or impossible today to reconstruct the find context of most of the tens of thousands of cuneiform texts excavated during the nineteenth century (Larsen, 1988).

The role of historical archaeology as a producer of texts is relatively limited today, perhaps with the exception of Chinese archaeology, Japanese archaeology, and Mexican archaeology. At the same time, excavated texts have attracted new interest with the increasing attention now paid to their find contexts. As the survey above showed, the stratigraphic context of the texts was registered in earnest from the interwar period onward. In the last few decades, more and more analyses have taken into consideration the complex association between the texts and their physical context, especially in the Mediterranean area, the Middle East, China, and Mexico.

Another expression of the philological tradition is the role of archaeology in providing background knowledge or realia to facilitate textual analysis. Linguists have long been aware that the interpretation of a text does not just build on the text itself but also presupposes a general knowledge of the context in which the text is created. For the older written sources in particular, studies of material culture provide a "preunderstanding" of and a background to the interpretation of the texts. This aspect already exists in the antiquarian tradition of the Sung dynasty, in which studies of old artifacts became a part of the textual analysis. For Lu Dalin, who drew up the first catalogue of ancient bronze vessels in 1092, the explicit aim was "to search for the origins of [ritual] institutions, to fill the lacunae in the classics and their commentaries, and to correct errors of the [classical] scholars of the past" (cited from Chang, 1986:9).

In Europe the realia aspect first appears during the Renaissance, when scholars began to study material remains of antiquity in order to solve linguistic problems. The classical meaning of many Latin terms was unclear, so a knowledge of the realia of life in ancient Rome was needed. An example of the use of ancient objects to solve philological problems is a ground-breaking numismatic study from 1515 by the French humanist Guillaume Budé (1468–1540). By means of a systematic investigation of ancient coins in the coin cabinets of his day, he succeeded in clarifying the monetary terminology and the value of the coins in antiquity (Weiss, 1969:167 ff.).

With the establishment of modern philology in the first half of the nineteenth century, there was a special emphasis on realia. Studies of material culture provided general background knowledge or a hermeneutic preunderstanding, which could also be used to solve linguistic problems or explain the meaning of individual words in an extinct language. The relation between *Sachen und Wörter*, that is to say, the relation between the things and the terms by which they are designated, was fundamental to Jacob Grimm (1785–1863), one of the pioneers of comparative linguistics (Platzack, 1979:123 ff.). This association meant that the things gave "flesh and form" to the words (Deshpande, 1967:453).

The link between textual criticism and archaeology is particularly clear in the growth of biblical archaeology, which was geared to a single, but crucial, text in the European tradition (see Figure 11). Yet the philological need for realia knowledge has also set its stamp on classical archaeology, Egyptology, Mesopotamian archaeology, and Indian archaeology. The realia aspect is especially clear in connection with the establishment of classical archaeology. As we saw above, Eduard Gerhard claimed in the mid-nineteenth century that the subject could only become a professional discipline when it was defined as "philological archaeology," and for Bianchi Bandinelli this was a valid description of the subject right up to the First World War.

Today the role of archaeology in providing realia knowledge is very small, although Chinese archaeology, with its association with the indigenous tradition of learning, has recently been criticized for being "an accessory to textual criticism" (An, 1989:17). As a monument to the age when many historical archaeologies served as ancillary sciences to philology, however, we still have the great dictionaries of the extinct historical languages. Thanks to the realia function, studies of material culture are still indirectly present in the linguistic interpretations given in these dictionaries, which are old, though still frequently consulted.

The third philological function of archaeology is its role as a reference point for historical linguistics. The establishment of modern philology meant that the biblical view of language was replaced by the comparative history of language. Instead of a divinely given ranking of tongues, historically related language families were established, such as the Indo-European and the Semitic languages (Bernal, 1987:224 ff.; Platzack, 1979:133 ff.). With this shift of perspective, new philological problems were formulated in the first half of the nineteenth century, and philologists soon turned to archaeology for help. This function as a reference point for historical linguistics is in most cases a secondary addition to the historical archaeologies. The exceptions are the postcolonial archaeologies in Africa and Oceania, which have been established in close association with historical linguistics (cf. Ehret and Posnansky, 1982; Kirch, 1986).

In comparative linguistics it is crucial to be able to date different horizons in the history of languages and to determine the geographic basis of these horizons. To reconstruct phases of a language going back beyond the first written evidence, archaeology has come to play an important role as a reference point in time and space. By means of archaeology it has been possible to discuss the "origin" and "dissemination" of different languages, for example, where the Indo-European languages came from and in what periods they spread over Eurasia (Figure 30). This dialogue between historical linguistics and archaeology has especially concerned the prehistoric archaeologies, since the problems circle around the reconstructed phases of the languages, before the oldest textual evidence. The debate about Indo-European has therefore mainly taken place outside the field of historical archaeology, although the methodological problems of relating language and material culture are identical with the problems of relating artifact and text (Mallory, 1989; Renfrew, 1987; see also Ehret and Posnansky, 1982).

A problem in historical linguistics that is more closely linked to the field of historical archaeology concerns place-names. The meaning and chronology of place-names have largely been studied on the basis of the appearance and history of the named places. In Europe, place-name types associated primarily with early medieval settlement have been

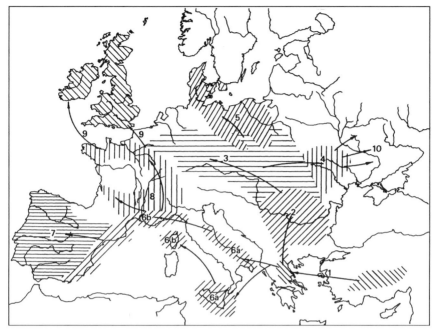

Figure 30. Hypothetical spread of the Indo-European languages in Europe during the Early Neolithic (Renfrew, 1987, fig. 7.7, by courtesy of Colin Renfrew). Colin Renfrew's controversial thesis that the Indo-European languages spread over Europe as the language of farmers in the Early Neolithic is an illustrative example of how archaeology can serve as a philological reference point.

studied in relation to archaeology. By comparing settlements and graves with different types of place-names, the dating of place-names in terms of historical linguistics has been studied and critically scrutinized (e.g., Copley, 1986; Wainwright, 1962).

To sum up the philological tradition, I would particularly emphasize that the material remains play a much more subordinate role in this tradition than in the aesthetic tradition. The starting points for studies and analyses have been and continue to be linguistic problems. There is however a possibility of a more mutual relation in the relatively new interest in the spatial context of the texts. At the same time, the role of archaeology as an ancillary subject to philology—especially as realia knowledge—means that historical linguistic knowledge is fundamentally dependent on studies of material culture. Historical archaeology thus appears here less as a "text-aided archaeology" and more as an "object-aided philology."

Extending the Text: The Historical Tradition

The study of material culture as a way to extend text-based history is the most widespread tradition in the field of historical archaeology today. This historical trend is primarily geared to topography, protohistory, and questions of technology, economy, and social conditions. The tradition can be detected in several subjects in the nineteenth century and the early twentieth century, but the real breakthrough came after the Second World War. The perspective is directly connected with the establishment of medieval archaeology, Japanese archaeology, and African archaeology, and with the emergence of landscape archaeology in several of the older disciplines.

It is a general human characteristic to use objects as support for the memory and as a starting point for narration. The systematic use of material traces of the past for historical narrative is known from European, Chinese, Japanese, and Arabic historiography. Herodotus explicitly based his history on "events" and "monuments," and Thucydides has been described as the first archaeologist, since he used objects in a critical way in his historiography (Hedrick, 1995). In Europe the role of material culture in the writing of history became particularly distinct with the antiquarian tradition of the Renaissance. An example of an author who systematically used ancient ruins, sculptures, and coins to write history was the Italian humanist Flavio Biondo (1392–1463) (Weiss, 1969:59 ff.). Through the antiquarian tradition, studies of artifacts remained closely associated with historiography until the establishment of modern archaeology. The fundamental feature of the premodern antiquarian tradition was that artifacts were viewed as illustrations of human history, which was considered to be known in its entirety through texts (see Trigger, 1989:70 ff.).

It was only with the establishment of the modern discipline of history in the first half of the nineteenth century that a more exclusively text-based scholarship was created. History was thus created in opposition to, and with the mutual exclusion of, the object-centered archaeology established at the same time. History in the nineteenth century focused primarily on political history. This historiography of state idealism centered on great events, ideas, and personalities (Burke, 1991:2 ff.). Because of this perspective, the "historical" role of archaeology in the nineteenth century was chiefly confined to studies of the historical topography that was the scene of the political drama, and of "early" history, which to varying extents lacked the "necessary" texts (see Figure 3). It was only when historians began to turn their attention to social, economic, and technological matters in the twentieth century that these problems also attracted archaeological attention.

Historical topography was a central problem in the nineteenth century that led directly to the emergence of subjects such as classical archaeology, biblical archaeology, Mesopotamian archaeology, and Indian archaeology. Historical topography was a form of historical background knowledge that provided the external contours of the political course of events. The texts spoke of landscapes, places, and monuments, but their location and appearance were often unknown. It was only by archaeological studies that the political arena could also be visualized. The great surveys of historical topography were often created in the initial phases of the history of the different disciplines. The greatest need for surveys was found in the areas that lacked direct continuity with older vanished settlement. This applied to countries such as Greece, Palestine, and India, where men like Leake, Robinson, and Cunningham managed to compile historical topographical surveys thanks to many years of studying terrain and text (see Figure 13).

In areas with greater continuity between the past and the present, studies of historical topography were confined to spatial details. Since most medieval cities in Europe still exist, topographical studies have not concerned the location of the cities but rather details in them, such as the location of demolished churches and the courses of vanished streets. Studies of historical topography still play a role in the historical archaeologies (see Figure 38). However, with the heavy emphasis on social and economic questions, the topographical interest has changed character. Topography is no longer seen as a physical frame for political events; it is instead seen as an expression of, for instance, agrarian and urban production. The boundary with settlement archaeology is thus quite diffuse.

Besides historical topography, "protohistory" was an accepted field of archaeological study in the nineteenth century, chiefly in classical and Indian archaeology. Early "historical" periods in Greece and India were known only from oral tradition that had been put into writing, as in the *Iliad* and the *Rigveda*, or from "ethnographic" descriptions by outsiders. Archaeology functioned then as an extension of history by complementing the vague information of written sources. The purpose was to write history as the term was understood then, as political and religious history, often with the focus on ethnic groups.

The concept of protohistory or early history has been criticized for being Eurocentric (Champion, 1985), since it presupposes that there are people without history. Yet the very approach of using archaeology to study periods with few contemporary written sources has remained an important tradition throughout the twentieth century. Many historical archaeologies are in fact geared to "early" historical phases, whereas later, "better-known" periods are left to historians of art and architecture. Postmedieval archaeology is very weakly represented in Europe, and many medieval archaeologists are not even interested in the later Middle Ages. In the eastern Mediterranean and the Middle East, the politically and culturally important Turkish period has received virtually no attention in archaeology (see Silberman, 1989). Indian, Chinese, and Japanese archaeology are mainly concerned with early phases in the history of the different countries (Figure 31; cf. Figure 15). In Latin America the Spanish conquest normally marks the chronological limit to archaeological work (but see Schaedel, 1992).

Since the nineteenth century, the protohistorical current has preserved the focus on political and religious history. A great deal of attention has been devoted to the often politically loaded issue of ethnicity, especially for early historical periods in Europe, the Middle East, India, Japan, and Central America. Following the lead of Gustav Kossinna, archaeological "cultures" have often been classed as ethnic groups (Trigger, 1989:163 ff.), but in recent years ethnicity has attracted a renewed, more critical interest. There has been particular discussion of the important but difficult question of whether ethnicity can be detected in material remains (Olsen and Kobyliński, 1991). After the Second World War, however, the interest in protohistory, like the study of historical topography, has been widened, so that today it also comprises matters such as production, distribution, and social organization (e.g., Hodges, 1982a; Morris, 1987; Randsborg, 1980).

From the beginning of the twentieth century, the approach of "extending the text" took on a new meaning, as historians broadened their interest from political events to social, economic, and technological matters. Especially after the First World War, political events began to be shifted into the background, giving way to studies of topics such as demography, agriculture, craft production, and trade. The concentration on these questions also meant that certain historians began to use sources other than written documents. An example is the *Annales* historian Marc Bloch (1886–1944), who studied, among other things, agricultural tools in order to write the history of French farming (Nordenstam, 1993:223). This shift of perspective can be detected not only in the first generation of *Annales* historians but also in the emergence of special subjects such as economic and social history and in the establishment of Marxist historiography in the Soviet Union (cf. Burke, 1991; Stoianovich, 1976:25 ff.; Trigger, 1989:216 ff.).

Because archaeology was viewed as part of history in the Soviet Union, issues of production and power over production became particularly prominent in Russian archaeology. Through Gordon Childe, as is known, this early form of Marxist archaeology was passed on to

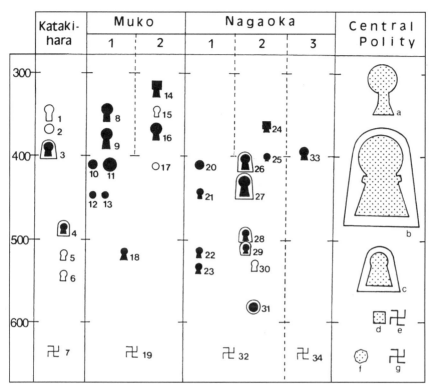

Figure 31. Chronological survey of large burial mounds (1–6, 8–18, 20–31, a, b, c, d, f) and Buddhist temples (swastikas) in three provincial areas on the River Katsura and in the central political area of Japan (Tsude, 1990, fig. 2, by courtesy of Hiroshi Tsude and Antiquity Publications). The figure shows how the formation of the Japanese state meant a change in practice: whereas the elite had formerly distinguished themselves by means of large burial mounds, often keyhole-shaped, they started to manifest themselves through Buddhist temples (see fig. 18). This study is an example of the strong protohistoric current in Japanese archaeology, as in many other historical archaeologies (see e.g., Morris, 1987; Randsborg, 1980).

prehistoric archaeology (Trigger, 1989:254 ff.). In the historical archaeologies, however, the influences mainly came from non-Marxist social and economic history. An example of this inspiration is the emphasis on market trade and urban craft production in the incipient medieval archaeology during the interwar years (Andrén, 1996).

It was not until the end of the Second World War, however, that the historical tradition came to dominate the field of historical archaeology totally. Historians and archaeologists alike have repeatedly emphasized the significance of archaeology in studies of, for example, economic and social conditions. Never before have entire landscapes, profane settlement, and different forms of production been subject to as much archaeological study as in the last 30 or 40 years (Figure 32). Intensive landscape surveys have often enabled a completely unknown settlement history to be reconstructed in areas all over the world (see Figure 4). Major excavations in abandoned or still existing historical towns have led to extensive archaeological debate about urbanization as a more or less global phenomenon (see Figure 8). At the same time, there has been great interest in preindustrial craft production. An area that has attracted particular attention is iron technology, such as questions of the spread of iron production in India and Africa (see Figure 20), or problems concerning the emergence and scope of mining in medieval Europe. Paradoxically, the breakthrough of the historical tradition in the postwar era has also led to a greater emphasis on science in archaeology. The discussion of topics such as production and demography from archaeological perspectives is possible on the basis of pollen analyses, macrofossils, osteological study of human and animal bones, and other evidence (see e.g., Hesse, 1995).

This historical tradition—with a focus on production and its social and technological determinants—is part of a general materialist trend in both history and archaeology in the postwar era. Materialism, which may be either Marxist or non-Marxist, has drawn attention to the fundamental agrarian economy in all preindustrial societies. In archaeology the materialist perspective has furthermore led to a special methodological approach. Economic and social conditions are viewed not just as important questions but also as particularly suitable objects for archaeological study, whereas politics and religion are considered more difficult to study (Trigger, 1989:266 ff.; see also Bernal, 1962; Jankuhn, 1973; Jansen, 1984). The best-known expression of this attitude in archaeology is Christopher Hawkes's (1954) "ladder of reliability," where technology is on the lowest and safest rung, and religion is on the highest, least secure rung.

This methodological outlook has been further emphasized in the field of historical archaeology with reference to the incomplete nature of the written sources. Since most early texts were written by and for an elite, the written sources often lack information about subjects such as agriculture and urban crafts. Archaeology is therefore seen not just as most suitable for the study of these matters, but also as the only way to investigate the "nonwriting" spheres of literate societies. In recent years there has been a special emphasis on the role of archaeology in writing the history of the underclasses or the people without writing (e.g., Hall, 1993; Orser, 1996:159 ff.).

In recent decades, however, the approach of "expanding the text"

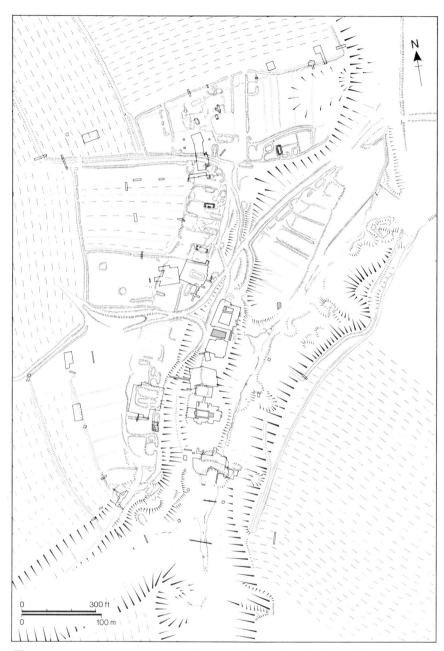

N

| 0 | 300 ft |
| 0 | 100 m |

Figure 32. Plan of the surviving earthworks of the medieval village of Wharram Percy in Yorkshire (Illustration by Chris Philo for Wharram Research Project, see Beresford and Hurst, 1990, fig. 2). The investigation of Wharram Percy (1948–88) represents one of the longest and most systematic excavations of agrarian settlement in Europe, as well as marking the pronounced orientation of archaeology toward economic and social questions after the Second World War.

has once again begun to acquire a new meaning, in conjunction with the new orientation of history toward anthropology (Burke, 1991:2 ff.). Areas that had formerly received little attention from historians, such as childhood, death, gestures, the body, and silence, have all had their histories written. Textual analysis is still the central part of the historian's work, but because of the new research fields, many historians have also begun to use pictures, oral tradition, and in some cases material culture. The archaeological counterpart to this historical reorientation is primarily seen in studies of "everyday life" in European medieval archaeology. However, since archaeological studies of everyday life have an older background in cultural history, I will consider them in the next section.

Summing up the historical tradition, it can be described as a way to write history archaeologically for periods and areas with few texts. This focus on zones without writing means that the complementary role of archaeology is emphasized above all. Archaeologists search more or less openly for gaps and defects in the written sources and fill them in with archaeological studies. Often all available sources—including texts—are used in the studies, but it is still ultimately the insufficiency of the written records that justifies the archaeologist's presence. In this complementary outlook there is a paradoxical similarity between historians and archaeologists, although they superficially appear to have diametrically opposed views. As I have mentioned, Moses Finley and other historians have seen the role of archaeology as being in inverse proportion to the availability of written sources, in a kind of zero-sum game (Morris, 1994:39). This attitude has been criticized by many archaeologists, but the actual research situation does bear up Finley's words. Even archaeologists who have vehemently asserted the intrinsic value and independent role of archaeology (e.g., Christophersen, 1979; Hodges, 1982a; Morris, 1987) work with protohistory or with economic and social questions. By virtue of their concentration on spheres more or less without writing, they conform through their practice to a tradition in which archaeology is seen as a complement to texts and text-based historiography.

Writing the History of Artifacts: The Tradition of Cultural History

The tradition of cultural history plays a less prominent role in the field of historical archaeology. This current is chiefly associated with European medieval archaeology and historical archaeology in the United States, but it has now received greater general attention as a result of contextual archaeology with its emphasis on the significance of artifacts.

The background to the tradition of cultural history may be sought in the anthropological and ethnological interest in foreign objects. Ethnographic descriptions of foreign peoples are known, for example, from ancient Greece, the Roman Empire, China, and the Arab world. These accounts often stress the divergent features of alien people by describing their material culture, such as building traditions, costumes, and tools. Yet these reports on unusual "barbarians" and their artifacts did not just show wonder about foreign habits but also allowed authors to reflect about their own culture. Descriptions of barbarians could thus serve as an internal critique of civilization. A well-known example is Tacitus's *Germania*, from A.D. 98, in which unspoiled Germans are contrasted with decadent Romans (Lund, 1993).

Foreign peoples became a common subject of description and debate in Europe with the global expansion of trade, starting at the end of the fifteenth century. Besides ethnographic descriptions in texts, objects brought home from foreign climes were also used as concrete illustrations of foreign cultures. In the seventeenth and eighteenth centuries, ethnographic accounts were used in serious attempts to write the history of civilization. For example, the narrative of Captain Cook's voyages in Oceania inspired the idea of the original social contract in political philosophy (Kirch, 1986). As in classical antiquity, foreign cultures were used as weapons in an internal critique of civilization, since they could be seen as superior to Europe. In a case like China, moreover, the advanced state of the country could be confirmed in luxury articles imported to Europe.

From the end of the eighteenth century, ethnographic knowledge was increasingly systematized, and simultaneously the "foreign" people in Europe—the timeless common people, or folk—attracted increasing attention. The twin interest in foreign people is clear in Johann Gottfried von Herder (1744–1803), who recorded and published German *Volkslieder*, and also wrote the history of civilization from an evolutionary perspective. This twin interest in foreign people, inside and outside Europe, led in the nineteenth century to two partly different traditions: on the one hand, cultural history or folklife studies/ethnology, which studied the peasantry in Europe, and on the other hand, ethnography/anthropology, which studied non-European societies, chiefly those without writing. The common feature of these foreign peoples was that they more or less lacked writing. Both the peasantry and the "primitives" could be studied primarily through their artifacts and their oral traditions. The importance of artifact studies was also marked by the new museums of ethnography and cultural history, in which artifacts from overseas colonies or preindustrial European folk cultures were collected. The focus on objects and oral traditions also meant that anthropology and ethnology considered different kinds of questions from those studied by the historian. The given object of study was not political events dictated by "great men," but more anonymous cultural categories, such as myths, kinship systems, housing, and costume (Harris, 1968; Hautala, 1971; Strömbäck, 1971; Svensson, 1966).

From the perspective of historical archaeology, it is, above all, the artifact-oriented tradition of cultural history that is of interest. In Europe the material folk culture of the seventeenth, eighteenth, and nineteenth centuries was studied, even though these objects came from a time with an overwhelming wealth of written sources. The basic attitude, however, was that it was possible to write a different history based on artifact studies (Svensson, 1966). Occasional instances of this artifact-based tradition of cultural history can be found in nineteenth-century medieval and Renaissance studies. An early example is the Swiss historian Jacob Burckhardt (1818–1897), who studied art and architecture in order to write a cultural history of the Italian Renaissance (1860), focusing on trends instead of events (Burke 1991:8 ff.). Inspired by Burckhardt as well as ethnology, the Swedish archaeologist Hans Hildebrand (1842–1913) looked for a compromise between the state idealism of his day and social history (Figure 33). Hildebrand, who probably coined the term "material culture," argued in a programmatic article from 1882 that archaeology in historical periods should be pursued from the angle of anthropology/cultural history, which meant that all aspects of the past could be studied, including political history. Besides "material culture," written sources could also be used, but "for a different purpose and in a different way" from that followed by contemporary historians (Hildebrand, 1882:21). On the basis of this program, Hildebrand (1879–1903) wrote a 3,000-page work on the cultural history of Sweden in the Middle Ages.

Hildebrand, however, was relatively alone among archaeologists in taking this stance, and the interest in material culture in Europe contin-

———————————————————————————————→

Figure 33. Detail of the sculpture of St. George from 1489 in Stockholm, reproduced to illustrate a discussion of the appearance and significance of chivalric equipment in the Late Middle Ages (Hildebrand, 1879–1903 (II), fig. 121). The picture is an example of a cultural historian's critique of the state-idealism that dominated historiography in the nineteenth century: "The [battle] dress determines the movements and the manner of fighting, and this in turn influences tactics, and resulting from all this is the politics based on war" (Hildebrand, 1882:26 f.).

ued to be associated with postmedieval folk culture and the discipline of ethnology, which was professionalized in the late nineteenth and early twentieth centuries. In the last 10 years, however, questions of cultural history have once again moved toward the center of medieval archaeology, partly as a result of the inspiration of anthropologically oriented history (Burke, 1991). Examples are the interest in everyday life in German and Central European medieval archaeology (Felgenhauer-Schmiedt, 1993; Meyer, 1985; Seidenspinner, 1989), as well as the interest in power, ideology, and everyday practice, which has also been inspired by contextual archaeology (Austin, 1990). A distinctive feature of this new interest in cultural history is an emphasis on the intrinsic value of material culture, which means that it is not necessary to look for lacunae in written documentation. An example is Matthew Johnson's study of dwelling houses in southern England in the sixteenth and seventeenth centuries. He is able to detect a change in the design of the buildings, from open interiors to houses divided into rooms, and he argues that the more closed houses are associated with Puritan ideology. Inspired by Anthony Giddens, Johnson claims that this change in the design of houses reflects an explicit practical knowledge or "practical consciousness" that precedes and partly creates Puritanism as a formulated ideology or "discursive consciousness" (Johnson, 1993:178 ff.).

Unlike in Europe, archaeology and historical archaeology in North America has become linked more directly to anthropology, which is a result of the colonial perspective on the continent. As mentioned above, Trigger maintains that archaeologists in Europe studied their own ancestors, so prehistoric archaeology was perceived as an extension of history, whereas archaeologists in North America studied the ancestors of the "foreign" aboriginal population, so archaeology became a part of anthropology (Trigger, 1989:316 ff.). Although historical archaeology has its origin in this North American anthropological tradition, restoration projects and the link with American folk studies have been decisive for the profile of the subject.

The tradition of cultural history is particularly clear in James Deetz (1977), who summed up colonial archaeology in the New England states. Inspired by Henry Glassie's folk study of vernacular architecture, Deetz has written a form of historical ethnography. On the basis of such different artifacts as pottery, houses, and gravestones, he has defined different periods in the cultural history of colonial settlement in the present-day eastern United States. Like medieval archaeologists oriented to cultural history, many historical archaeologists are working actively with artifacts, texts, and pictures (e.g., Yentsch, 1994). In this context Beaudry has repeated Hildebrand's more than century-old plea for an active archaeological reading of texts: "Historical archaeologists must develop an approach toward documentary analysis that is uniquely their own" (Beaudry, 1989:1).

The boundary between the tradition of cultural history and the historical tradition is sometimes diffuse, not least in today's convergence toward historical anthropology. A crucial difference between these currents, however, is the different methodological attitudes to written sources. From the point of view of cultural history, the presence or absence of texts makes little difference. All contexts with some form of physical expression can be studied, which means that archaeology does not have the same complementary role as in studies of zones without writing. In this tradition, archaeology is not searching for lacunae in the written sources but instead welcomes writing, preferably in deliberate confrontation with the texts.

Searching for Analogies: The Archaeological Tradition

A special tradition is linked to prehistoric archaeology and its need for analogies. This archaeological tradition, which has a distinct methodological profile, has never dominated the field of historical archaeology but can rather be seen as a possible aspect of all historical archaeologies.

Prehistoric archaeology is fundamentally dependent on analogies. The very establishment of the subject was based on a crucial analogy drawn with the "primitive" peoples described in ethnography (Charlton, 1981). Thanks to information about "primitive" people outside Europe, the "primitive" people of a past Europe were also discovered. This fundamental need for analogies is clear in Sven Nilsson (1787–1883), who formulated the very idea of a "prehistory," while simultaneously carrying out the first systematic ethnographic comparisons, in order to reconstruct prehistoric economy and technology (Stjernquist, 1983; Welinder, 1991).

Anthropological analogies have remained an important, but problematic, element in prehistoric archaeology ever since Nilsson's days. On the other hand, it is only in the last 30 years that it has become common to draw analogies to societies with writing, as a way to study material culture in "controlled" environments. The aim of these analogies varies; it can be pedagogical, methodological, or theoretical. As a pedagogical aid, present-day material culture is used to explain archaeological methods at many universities (cf. Wilk and Schiffer, 1981). An early pedagogical example is Oscar Montelius's (1900) description of the development of the railway carriage, to explain typology as a method.

Analogy may also be a way to test archaeological methods, as has been maintained especially in the last 35 years (Bernal, 1962). The principle has been clearly formulated by David Clarke, who claims in his programmatic article "The Loss of Innocence" that amateurs, historical archaeologists, and field archaeologists are perhaps the groups that are least interested in new archaeology, but that "work in text-aided contexts will increasingly provide vital experiments in which purely archaeological data may be controlled by documentary data, bearing in mind the inherent biases of both" (Clarke, 1971:18). One of the best-known experimental investigations is James Deetz and Erwin Dethlefsen's study of gravestones in Massachusetts (Figure 34). By examining 25,000 gravestones from the eighteenth and nineteenth centuries, they were able to test—and confirm—seriation as a method and to discuss the significance of the course of innovation for chronological systems, the so-called Doppler effect (Deetz and Dethlefsen, 1965; Dethlefsen and Deetz, 1966).

Finally, analogy is used to develop theories about the relation between human action and material culture. In its most systematic form this trend is called "historical ethnoarchaeology" (Welinder, 1994), and it is a branch of general ethnoarchaeology that has been developed in processual and postprocessual archaeology (Gould, 1978, 1980; Hodder, 1982; Miller, 1985). An example of this genre is Stig Welinder's (1992a, 1992b, 1994) studies of a nineteenth-century village in central Sweden. By means of archaeological excavations, studies of written sources, and interviews, he has developed theories about material culture that have since been applied in purely archaeological contexts. Even modern society, with its high density of texts, has been the subject of ethnoarchaeological research, such as the studies of waste management in today's U.S. (Figure 26) (Rahtje and Murphy, 1992). Another example is the study of English and Swedish beer cans by Michael Shanks and Christopher Tilley (1987:172 ff.). By linking variations in the design of beer cans to alcohol consumption and alcohol policies in the two countries, they are able to show the complexity of material culture and its active role in the view of alcohol.

The function of the historical archaeologies as a text-controlled archaeological laboratory has been stressed in recent years as an extra argument for this work (e.g., McGuire and Paynter, 1991:19; Morris, 1994:45). Yet there has also been extensive criticism of this methodological tradition. Thomas C. Charlton (1981) calls for arguments to show that the results of the "tests" are also relevant for prehistoric times, and C. J. Arnold (1986) thinks that the very idea of text-controlled tests is naive, since it builds on an overestimation of the evidential value

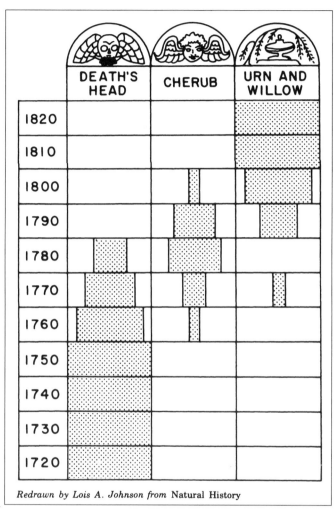

Figure 34. Seriation of stylistic elements on gravestones in Stoneham, Massachusetts (Schuyler, 1978:84, after Deetz and Dethlefson, 1967, fig. 1, by courtesy of Baywood). James Deetz and Edwin Dethlefson's study of gravestones from the eighteenth and nineteenth centuries in Massachusetts is a classic example of how historical archaeology can be used as a laboratory for testing archaeological methods. Since the gravestones are dated, the relative distribution of stylistic elements shows that seriation is a chronologically valid method.

of written sources. For Mary C. Beaudry (1989:1), the purely methodological profile of model-testing means that "the questions answered
through such studies bear little interest for serious students of the New
World's past." Barbara J. Little (1994:12) asserts that "instead of truly
acting as a laboratory, historical archaeology often offers itself only as
a confirmation of models already created and applied to other data."

To sum up the archaeological tradition, I would say that the search
for historical analogies is a search for many and ample written sources.
The starting point for this tradition is thus the exact opposite of the
historical approach, which applies archaeology where texts are few or
nonexistent. Since the tradition is more purely methodological than
the other currents, it has never dominated the historical archaeologies.
It is rather an aspect that is recurrently cited as one of several purposes
of the historical archaeologies. However, the criticism in recent years
of this methodological approach shows that the current debate about
analogies in prehistoric archaeology must be integrated in the tradition
if it is to be anything more than a rhetorical defense of the historical
archaeologies.

Summing Up

As in the global survey of the previous three chapters, the historical archaeologies can be described as separate disciplines studying
different parts of the human past. By concentrating on the shared
methodological problems, however, it is also possible to describe the
different subjects as a coherent historical archaeological field. I have
described this area in a new way by presenting five different approaches
or traditions that can be detected in most of the disciplines. The five
traditions are presented separately, with some of their characteristic
features highlighted. In practice the contours are often less clear, and
the boundaries between the traditions are fluid. The philological and
archaeological traditions, which are most obviously oriented to methodological problems, are often found as aspects of broader scholarly
work. The boundary between the traditions of history and cultural
history is in some cases obscure, especially with the growth of historical anthropology and with the use of anthropological theory in historical archaeologies (see e.g., Hodges, 1982a; Morris, 1987; Renfrew, 1972).
At the same time, there are distinctions between the different traditions
that it may be important to stress for various reasons.

The internal discussions of the "true character" of different historical archaeologies can be made more comprehensible if they are
seen in relation to the five traditions and their varying significance. In

Africa and Oceania the historical archaeologies have had a clear historical profile ever since they were established in the 1960s. Archaeology
has been used to write noncolonial history, as a contrast to colonial
anthropology's perception of these areas as timeless, stagnant continents "without history." In classical archaeology too, the historical tradition has become distinct, but in this case it can instead be seen as a reaction to the former dominance of the aesthetic tradition. Instead of
studying the aspects of antiquity that are known from written sources
and have served as exemplary models, archaeology has turned its attention to the more anonymous sides of the classical world, such as agrarian
settlement. In medieval archaeology, by contrast, which has by tradition been very closely linked to the discipline of history, there is explicit
skepticism today about viewing archaeology as a complement to written
sources. Some medieval archaeologists have instead advocated the perspectives of aesthetics and cultural history (cf. Austin, 1990; Gilchrist,
1994; Moreland, 1991).

   In the same way, different stances on the relation between material
culture and writing can be grouped in accordance with the five traditions. In the aesthetic tradition the difference between artifact and
text is important. Texts are used where they exist, but they are considered insufficient for bringing the past to life. The philological tradition
emphasizes the similarity between material culture and writing, since
it seeks identity between artifact and text in order to solve linguistic
problems. In the historical tradition too, this similarity is emphasized,
since artifacts are studied to complement the written sources. In contrast, the tradition of cultural history seizes on differences between
artifact and text, since it stresses the intrinsic value of material culture. Finally, a basic similarity is assumed in the archaeological tradition, since the quest for analogies must build on correspondences between artifact and text. It is important to be aware of this link between
the five traditions and the view of the relation between artifact and text
if one wishes to understand the field of historical archaeology and its
specific problems. After the following sketch of historical archaeology in
a general historiographical context, I shall return to this link and its
methodological consequences.

THE FIELD AS MODERN DISCOURSE

   The field of historical archaeology can be described as consisting of
distinct, specialized subjects and as boundary-crossing methodological
approaches, but also as an intermediate zone between more object-

based and text-based subjects. Since this dilemma of "in-betweenness" is a result of the specialization of modern science, it is important to view the field of historical archaeology in a broader historiographical context. It is also important, however, to deal briefly with this specialization in relation to the political and ideological functions of scholarship.

The history of modern science has been well told many times before (e.g., Bredsdorff et al., 1979; Kuhn, 1970; Liedman, 1977; Nordenstam, 1993; Toulmin and Goodfield, 1965), so I shall just hint at some historical archaeological aspects of this history. To put it very concisely, modern science can be seen as a secular cosmology that in the first half of the nineteenth century replaced the biblical cosmology that had dominated in Europe. In the eighteenth century, nature and humanity had been seen primarily as part of "the great chain of being," as a divinely given, static hierarchy (Lovejoy, 1936). In the first half of the nineteenth century, however, this great chain of being was "temporalized" as a result of empirical studies of increasingly specialized subjects. The earth, plants, animals, and humans and their languages received a long, complex history, without divine intentions. The world was viewed in an evolutionary perspective, and the search for origins became one of the most important bases of explanation. Yet this was an evolutionism with many nuances. Development could be seen as unilinear or multilinear, as goal-directed or non-goal-directed (Toulmin and Goodfield, 1965).

When Thomsen and Rickman in the 1810s began to use typological and stratigraphic methods, in order to create chronologies for material culture, they personified just two of the many expressions of an evolutionary way of thinking that had its breakthrough in Europe between 1810 and 1830 (Toulmin and Goodfield, 1965:233 ff.). Parallel but mutually independent approaches show the difficulty of writing the history of a discipline. A disciplinary perspective often emphasizes the internal influences exerted by a few significant persons, but in a broader perspective it is often easy to see parallel ideas that are due to a shared intellectual climate, a "Zeitgeist."

Despite the shared evolutionary idea in nineteenth-century science, there were different types of breaches with biblical cosmology. The basic idea for prehistoric archaeology—as for geology, biology, and paleontology—was a breach with the biblical premise that the whole history of humanity and the world was known through texts. The creation story could no longer be used even as an analogy for interpreting the distant past or for calculating the inconceivable length of time before the oldest written sources. The fundamental task of these subjects was thus to write "history" without the aid of texts. This lack of written sources has ever since been the methodological point of departure in prehistoric archaeology. For example, it was the lack of texts from the North European Bronze Age that led Oscar Montelius (1843–1921) to develop his typological method for Bronze Age artifacts (see Toulmin and Goodfield, 1965:141 ff., 197 ff., 232 ff.; Trigger, 1989:73 ff., 155 ff.).

In subjects such as history, philology, and the history of law, the breach with biblical cosmology was different. The biblical perspective was instead relativized *by means of* texts. The amount of texts increased enormously with the collection and excavation of old texts, with the decipherment of unknown scripts (see Figure 35), and with the opening and ordering of formerly closed archives. At the same time, the view of texts changed radically as a result of philological textual criticism. Texts were critically scrutinized, questioned, and sometimes totally rejected. Source criticism became a crucial principle of text-based subjects. Only after such critical scrutiny was it possible, in the words of Leopold von Ranke (1795–1886), to write history "as it actually happened" (Burke, 1991:3 ff.; Nordenstam, 1993:77 ff.; Toulmin and Goodfield, 1965:235 ff.). Von Ranke's standpoint is today often critically regarded as the very creed of historical empiricism, but when these words were formulated in 1824, they were a radical plea against a simple and naive belief in texts.

In the historical archaeologies, which work with artifacts as well as texts, there are elements of both these breaches with biblical cosmology and its view of texts. On the one hand, most of the nineteenth-century studies in the history of style presuppose a knowledge of material culture "beyond" the texts. On the other hand, the presence of texts has been a driving force for both fieldwork and methodological development. As we have seen, the hunt for texts was an important incentive to archaeology in the Mediterranean area and the Middle East. At the same time, stratigraphic methods were furthered by the difficulty of relating texts to the often complex objects of study, such as monumental buildings and settlements with occupation layers reaching a depth of tens of meters. Examples are Fiorelli's archaeological school in Pompeii in the 1860s and Dörpfeldt's development of stratigraphic methods at "Troy" and Olympia in the 1870s and 1880s.

The nineteenth-century evolutionary cosmology, as many scholars have pointed out, became a Eurocentric worldview, since it was created in Europe, which was beginning to dominate the rest of the world. With the aid of the new cosmology, the domination of the great European powers could be explained as "natural," as a result of, for example, race, cultural development, art, and history (e.g., Fabian, 1983; Kuper,

1988; Said, 1978; Young, 1990). In Europe, moreover, evolutionism had a nationalist stamp, since the perspective could be used to define the distinctive national character in historical terms.

Evolutionism and its ideological connotations are directly linked to the emergence of prehistoric archaeology (Trigger, 1984, 1989:110 ff.), and this is also very clear in connection with the historical archaeologies (Morris, 1994). In evolutionism, civilization was defined as the existence of, among other things, writing and formalized art and architecture. Since the historical archaeologies study literate societies, they participated directly in the evolutionary categorization of the world. Historical archaeology was pursued in the Mediterranean area, the Middle East, and India, but not in Africa and Oceania, since "dynamic," text-using civilizations were reserved for historical archaeologists, while stagnant, "primitive" cultures were the preserve of anthropologists, ethnologists, and prehistoric archaeologists (see Figure 19). It has often been pointed out that this evolutionary division of the world is still largely preserved in the museums of the Western world. European art tends to be physically separated from the artifacts in museums of ethnography, archaeology, and cultural history (e.g., Trigger, 1989:128).

Evolutionism acquired its most concrete form in classical archaeology and in the nascent medieval archaeology, which were both methodologically linked to the historicizing styles of art and architecture that dominated in Europe and the United States throughout the nineteenth century. In neoclassicism and later also neo-Gothic and neo-Romanesque, architects of the time could play with a whole range of forms with different associations and meanings. The historical styles were used above all for the new monumental buildings that were necessary for the "bourgeois public sphere" of the modern nation-states: parliaments, city halls, courts, churches (see Figure 5), universities, museums, theatres, opera houses, hospitals, railway stations, and banks. These styles represented a European aesthetic that saw ancient Greece as the origin of "fine art" (see Figure 2). This aesthetic was then given a national touch in each separate European country by means of an emphasis on the country's artistic heritage, not least the medieval monuments. The European concept of art and European art history thus became the yardstick by which the artifacts and "artistic expressions" of other cultures could be defined and evaluated (Mosquera, 1994). The Eurocentric aesthetic was given a special application when it was also transferred to racial doctrine, and Winckelmann's concept of beauty served as the starting point for assessing the value of non-European peoples and their physical appearance (Gould, 1981: 23 ff.).

The focus on early history in the historical archaeologies is likewise a typical evolutionary search for origins. From a Eurocentric point of view, "history" was reserved for civilizations with writing, and Europe was at the apex of development. The great European powers were therefore seen as the rightful heirs of early civilizations that had vanished. This idea, which legitimated Europe's political mastery of the world, went hand in hand with the methodological view that "early" historical periods with few written sources were more suitable for archaeological study than later periods with more texts. The colonial powers were more interested in shouldering the heritage of a magnificent past than studying the "primitive chaos" that they had conquered (Figures 7, 9, 13). The same methodological attitude can also be linked with European nationalism, since it was more important from a national point of view to study a distant golden age than later, less glorious, periods in the modern history of the nation-states.

The evolutionary cosmology held an optimistic view of progress, but by the end of the nineteenth century this was increasingly replaced by a cultural pessimism, partly brought on by the obvious social problems caused by the industrialism in the European states. In the human sciences, such as archaeology, the evolutionary idea was modified by diffusionism. Instead of presupposing continuous progress everywhere, the basic idea of diffusionism was that only people in certain places and certain circumstances were innovative (Trigger, 1989:148 ff.). A special form of archaeological diffusionism is represented by the preoccupation with culture groups, which were often identified in nationalist terms with ethnic groups. Archaeological studies of culture groups have been continued in many parts of the world for most of the twentieth century. Trigger (1989:174 ff.) has singled out Chinese, Japanese, and Mexican archaeology in particular for having carried on these diffusionist ideas, but they are also found in Eastern Europe, both before and after the fall of the Soviet Union (e.g., Kobyliński, 1991).

A more radical breach with the evolutionary idea is seen in the synchronic and, in a broad sense, functional perspectives that were established from 1900 onward (Nordenstam, 1993:197 ff.). Instead of the historical background, the emphasis was now on contemporary perspectives and functional links. These changed temporal perspectives were directly linked to a changed view of humanity. Instead of viewing humans as acting more or less on the basis of their own free will and reason, scholars pointed out how human actions were bound by given structures, such as the subconscious, language, society, and nature. This change in modern secular cosmology can be detected in such different expressions as functionalist architecture, modern art, logical positivism in philosophy, and sociology. Like evolutionism in the nineteenth century, the synchronic perspectives in the twentieth century have differed widely. In structuralism and early functionalism, historical changes were almost totally rejected. In other cases, historical change has been viewed as important but it has been explained in functional terms involving conflict to a greater or lesser extent, as in system theory and structural Marxism (Nordenstam, 1993:189 ff.).

As with evolutionism, there are both similarities and differences in the synchronic perspectives in prehistoric and historical archaeology. Prehistoric archaeology, which by definition studies societies without writing, has taken its inspiration from functionalism, structuralism, and Marxism in anthropology, often in the form of explicit anthropological theory (Trigger, 1989:244 ff.). In the historical archaeologies, on the other hand, the new synchronic outlook can often be seen in other contexts. In subjects that were methodologically tied to aesthetic approaches, the change of perspective was directly noticeable in the rejection of the historicizing styles and the change of restoration philosophy. Instead of reconstructing more or less imaginary original versions, the emphasis was now on the changing function of the buildings through time (see Figure 28). In contrast, when new functional questions concerning economic and social conditions attracted attention, the inspiration mainly came from historical studies of literate societies. In particular, historical works dealing with the same areas and periods as the various archaeological specialities have been sources of inspiration; an example is Henri Pirenne's studies of medieval cities (1925) in relation to medieval archaeology. The functional perspectives in many historical archaeologies have therefore started from questions and problems that were formulated in relation to specific historical periods rather than from general anthropological theory (Austin, 1990). It is because of this difference that many historical archaeologies are sometimes viewed—on a superficial level—as lacking theory.

These parallel but different sources of inspiration have meant that functional perspectives can be detected simultaneously in different areas. In several contexts I have already pointed out the global interest in landscape and settlement after the Second World War. Yet it is not enough to see this interest as a chain of stimuli starting from, say, American surveys in Peru and Mexico; it should instead be viewed as contemporary archaeological expressions of functionalist ideas in the postwar era (see Figures 4, 32).

Like evolutionism, the synchronic perspectives have ideological aspects. On the one hand, functionalism opened up for more relativistic perspectives. Societies could, in a sense, be regarded on their own

premises, without evolutionary comparisons. On the other hand, however, this preoccupation with functions and functional problems was
connected with the social problems that became obvious from the end of
the nineteenth century in Europe and its colonies. A politically important question was how social stability could be furthered, and this
problem is the focus of many early studies in functionalist sociology
and anthropology. Functionalism therefore became an intellectual current geared mainly to understanding and indirectly preserving the
power of the West over the rest of the world, albeit in reformed guise
(Klejnstrup-Jensen, 1978; Trigger, 1989:244 ff.).

Functionalism and its concepts, moreover, were based on European
society, which thus became a "natural" model for more global comparisons. Western society was no longer perceived as the highest stage of
human evolution but was held up as superior according to functional
criteria. In functionalist and structuralist perspectives too, the "primitive" Africa remained primitive, since much of the continent lacked
important criteria for civilization such as cities and writing (Kuper,
1988). Since historical archaeologies are primarily concerned with literate societies, the subjects have also helped to categorize the world on
the basis of functional perspectives. It is this use of Western concepts
in archaeology and historical archaeology that has been criticized from
an African perspective (Andah, 1995).

Modern science is a secular cosmology that has dominated the West
since the middle of the nineteenth century. For the last decade or so,
however, this discourse appears to have reached a new turning point:
postmodernism. This can be perceived in a great variety of ways, as
everything from a radical breach with modern science and its rational-
ism to a quickly passing academic fashion. I myself follow David Harvey (1989) in seeing the postmodern condition primarily as a new form
of modernism, but without Europe and the United States as the obvious
center. This is a modernism that expresses itself in radical pluralism
and lack of consensus, in a world where economic power is no longer
bound to a place or a nation-state, where encounters between different
people are more intensive than ever before, and where global media
communication breaks down the barriers of space (Harvey, 1989; Hemer, 1994; Young, 1990).

As in connection with the "synchronic" breach with evolutionism,
there will certainly be changes of scene in the scientific world, with
new disciplines forming and existing ones converging. The most important change in terms of method is that the empirical basis for the old
disciplinary boundaries is being dissolved. Historians are working with
images and artifacts in order to study more material aspects of the

past, such as the history of gestures and of the body (Burke, 1991). Archaeologists derive inspiration and analogies from philosophy and literary theory and have thus transgressed the methodological principle in prehistoric archaeology of working "outside" texts (see e.g., Tilley, 1990). And in the rapidly expanding field of interart studies, scholars are busy analyzing artistic expressions such as painting, literature, film, and music together (see Lagerroth et al., 1993). This area shows many similarities to the field of historical archaeology, since it concerns studies in which different forms of human expression are related to each other. Although the individual historical archaeological disciplines have in many cases existed for a long time, the extensive debate on artifacts and texts in recent decades, and the plea for a more coherent historical archaeology—such as the present work—may also be an expression of a form of boundary transgression that is typical of our age. The common denominator in the recent debate is the revolt against dependence on texts. This is part of a general relativizing of writing that has arisen because the text in our postmodern times has encountered competition from other media. As a result of the electronic revolution, such as radio and television, a secondary orality has arisen, which means that the historical construction of the literate mentality has received a whole new attention (Ong, 1982).

The postmodern condition also means a challenge to the Eurocentric perspective of modern science. More or less unspoken ideological stances in archaeology and other human sciences have been exposed in critical studies in recent years. I think that these critical perspectives are important, so I have hinted at them in my account. On the other hand, human science cannot be reduced solely to its ideological purpose, and knowledge cannot be completely relativized. The Indian archaeologist K. Paddayya (1995) has recently rejected some of the criticism of Eurocentrism, since he sees this as an academic stance in Europe and the United States, of no real interest for the world that is actually hit by Eurocentric perspectives. Paddayya argues that it is important to be aware of and to understand the British archaeologists' orientalist—and sometimes downright racist—view of India, but he also thinks that their work cannot be reduced to their ideological views (see Figure 13). Knowledge appears to be more two-edged than that, since the Indian liberation movement at the start of the twentieth century was able to use archaeological results as arguments against British power, and since a great deal of the work of British archaeologists is still valid, although the perspectives on India's past are different today.

In the last few years, several scholars (e.g. Champion, 1990; Orser, 1996:57 ff.) have argued that archaeology should be especially well suited to creating alternatives to the Eurocentric worldview, since material culture reflects a broader historical spectrum of the past than texts do. It may however be questioned whether artifacts have more critical potential than texts. Monumental architecture and formalized art are undoubtedly expressions of the elite (Kemp, 1989), and even more modest objects are often an expression of a dominating ideology. Martin Hall (1994) goes as far as to say that an invisible underclass can only be detected indirectly, in the contrasts between artifact and text. In addition, there is a critical potential in texts, which archaeologists do not always notice. Texts can be read critically (Beaudry et al., 1991), and texts from different contexts and traditions can be compared. For example, much of today's critique of civilization is based on Nietzsche, whose criticism of Europe was highly text-based, since it built on detailed knowledge of Persian literature and philosophy in particular.

The possibilities of the postmodern condition perhaps may rather be found in a renewed and unprejudiced encounter between European and non-European intellectual traditions, such as Indian and Chinese philosophy. This comparative philosophy could be especially interesting in the field of historical archaeology, since it comprises archaeologies from the whole world. As we have seen, archaeologists in India and Sri Lanka have already drawn attention to this perspective and begun to compare the view of the indeterminability of the world in postmodern theory with Hindu and Buddhist philosophy (Manatunga, 1994; Paddayya, 1990).

To sum up, I maintain that the field of historical archaeology can be perceived and described in several different ways at the same time. It can be seen as an area divided between strictly specialized subjects, and an intermediate zone crisscrossed by different, partly related traditions, and as a small part of modern human science. These perspectives do not exclude each other; they are just different levels in the definition of the field. In every subject one can discern one or more boundary-transgressing traditions, and in every tradition there are more or less clear expressions of the shared scientific currents of thought. These levels are also closely associated, since specialization is determined in large measure by the different transgressing traditions and the scientific discourse that dominates these traditions. For example, an evolutionary aesthetic and philological tradition went hand in hand with the establishment of classical archaeology, and a functionalist historical tradition can be linked to an important part of the renewal of the subject after the Second World War.

The crucial question, finally, is whether the field of historical archaeology—and the relation between artifact and text that defines this field—is also significant in a postmodern discourse. From a postmodern point of view, the opposition between material culture and writing could be seen as an uninteresting legacy of modern science and its empirically based specialization. But since I regard postmodernism as a form of modernism, I think that the difference will persist, although not all contemporary disciplines will necessarily survive. The survey of the historical archaeologies and their traditions shows, moreover, that the issue of the relation between artifact and text has been of interest outside modern human science, although it has been noticed particularly clearly in this discourse. Reflections of a debate about artifact and text can be found both in Renaissance Italy and in Sung-dynasty China. Moreover, in most literate societies in the last 5,000 years, writing has been perceived as a special phenomenon that has required explanations in mythological narratives (see Diringer, 1968:1 ff.). In other words, I think that the relation between artifact and text remains important in the modern discourse that is called postmodernism. This makes it essential to study in closer detail what this relationship consists in, and how a dialogue can be started between material culture and written sources. The next chapter will be about these methodological questions.

The Dialogue of Historical Archaeology

INTRODUCTION

The shared problem for all historical archaeologies is the encounter of material culture and writing. The question of what this dialogue is like and what artifact and text "really" are is ever present, more or less explicitly, in all methodological debate in the historical archaeologies. It is difficult to answer this question fully, but I try here to discuss it on the basis of the assembled experience in the field of historical archaeology.

A distinctive feature of this experience is the lack of consensus. Artifact and text can be perceived as both similar and different. From a textual perspective in particular, that is, from a viewpoint that artifacts should be seen as texts, the similarities have been stressed in recent years (see e.g., Christophersen, 1992; Wienberg, 1988). However, many other scholars have argued that artifact and text are in principle different. Using various metaphors, the differences have been described as different concepts (Wainwright, 1962:164 ff.), categories (McNeal, 1972), surfaces (Klejn, 1977), analogies (Andrén, 1988; Charlton, 1981), traces (Andrén, 1985:9 ff.), events (Snodgrass, 1985a), communication (Shanks and Tilley, 1987:85), cultures and processes (Leone and Crosby, 1987), traditions of abstraction (Netherley, 1988), and echoes (Murray and Walker, 1988).

Yet none of these stances is without problems, since the ultimate consequences of both perspectives are paradoxical. If artifact and text are seen as identical, then Finley's harshly criticized view is valid: The more written sources, the less the need for archaeology. This would mean that archaeological studies of periods rich in texts are only of methodological, "laboratory" interest. But if artifact and text are instead perceived as different, this reduces our possibilities for communication and translation between text and artifact. The question, then, is whether the historical archaeologies are caught between tautology and silence. I believe that this paradox can be solved by examining more analytically the relation between artifact and text and how this question has been handled in the historical archaeologies.

DEFINING PRACTICE

I do not see the definition of artifact and text as an end in itself, but as a way to understand different problems in the relation between material culture and writing and as a way to see how these problems can serve as a starting point for continued studies in the historical archaeologies. There are several paths to take in a more thorough study of the relation between artifact and text, for example, via philosophical studies of the concepts. In this work I have chosen a different course, namely, a historiographic and analytical investigation of the historical archaeologies. This choice arises from a desire to see how far experiences in archaeology, especially historical archaeology, can contribute to a general debate about material culture and written sources. It is thus largely an attempt to analyze practice.

To begin with, it is important to note that the relation between artifact and text is neither unambiguous nor static. Both artifact and text and the relation between them can be perceived in different ways, depending on diverse perspectives and traditions. We can see artifact and text as categories, as objects, as documents, or as discursive contexts, and in each of these perspectives the relations can be defined differently. The definitions of material culture and writing are thus contextual, and at least some of the conflicting stances are due to the fact that scholars are arguing from different perspectives.

As *categories*, artifact and text are relatively dissimilar. Writing is a strictly delimited category, since it is a representation of the spoken language. It is a form of technology that, in global terms, can probably be traced in only three courses of innovation. The alphabetic writing that predominates today represents just one course. Even though it is possible to follow the prehistory of writing for 5,000 years in Mesopotamia, the final cuneiform is nevertheless a qualitative leap. Writing is an invention or a "conceptual revolution" (see Harris, 1986:122 ff.), no matter how much its areas of use and social implications may have varied through time and place (Collins, 1995).

Material culture, in contrast, is an umbrella term for much more varied and heterogeneous things. Material culture represents a number of different artifact categories, each with its different history, uses, and meanings. Ian Hodder (1994) has recently questioned the concept of material culture precisely because it is such a cumbersome covering term. In a similar way, Barry Kemp (1989) has previously argued that the boundary between artifact and text is not always decisive, since an important dividing line runs between formalized and nonformalized objects. He sees architecture, art, and text in one context that differs from other categories such as tools and food refuse. I myself have stressed the textlike character of some material culture, one example being formalized architecture (Andrén, 1988). Despite some criticism of my ideas (Christophersen, 1992; Nordeide, 1989), I would still maintain that there is a difference between a refuse pit and a cathedral. The refuse pit may be a deliberate construction of great interest, for example, for studying spatial perceptions or attitudes to impurity. Yet the potential function and meaning of the refuse pit is more limited than that of the cathedral, and the mobilization of time and resources to dig a pit is completely different from the building of a cathedral.

As *objects*, however, artifacts and texts are identical, since all texts are artifacts. This material outlook on writing is important from several points of view. D. P. Dymond (1974:109 ff.) has pointed out that this perspective is fundamental to a great deal of work in diplomatics and epigraphy. The study of early texts in these specialities concerns matters such as the material on which texts are written and the form of the script. The object perspective may also explain why there are great similarities between historical and archaeological source criticism. In both cases, the source criticism may include the question of the degree of preservation of the source material. And in both cases the degree of preservation depends on human actions as well as natural processes in the course of the centuries. The material form of texts has also been used methodologically in large-scale historical studies. An example is Michael Clanchy's (1979) investigation of literacy in medieval England. He has studied wax consumption in the royal chancery to obtain a measure of the number of sealed documents and hence to estimate the extent of literacy.

From a social and economic perspective too, the materiality of the text is important. Writing is a form of technology and, like other technology, its meaning depends on who masters the technology, for what purposes it is used, and how it can be learned (Collins, 1995; Larsen, 1984). A recurrent pattern that clearly indicates the limited and elitist character of writing is that many early written languages were sacred languages that were no longer spoken, such as Sumerian, Sanskrit, rabbinical Hebrew, and Latin. Literacy in these cases was directly linked to a knowledge that was difficult to master and that took years of specialized instruction to learn (Ong, 1982). The elite character of writing is also seen in the fact that the materials on which texts were written were often expensive, not easily available products, such as papyrus, parchment, and silk. The manufacture of these materials was so important that text-producing institutions themselves often controlled this production. For example, many medieval monasteries secured a supply of parchment by breeding their own cattle.

The materiality of text has also been considered in a more fundamental way by some archaeologists. Both Stephen T. Driscoll (1988) and John Moreland (1991) maintain that the varying scope and meaning of literacy must first be determined before artifact and text can be compared with each other. Driscoll argues that our modern literate perspective forces on to the past a "natural" division into object and document. Instead of this difference, he advocates a combined analysis of artifact and text, since he refuses to distinguish artifact manufacture from writing.

As *documents*, or as cultural expressions, the character of artifact and text is debated. The question is what material culture and writing have to say about the past: Are texts "better" than artifacts at expressing certain phenomena, or vice versa? As is clear from the previous chapter, about the five boundary-crossing traditions, artifact and text can be perceived as both similar and different cultural expressions. The philological, historical, and archaeological traditions stress the similarities, whereas the differences are more important in the aesthetic and cultural historical traditions. In my opinion, these diverse outlooks are not necessarily incompatible but rather different aspects of the relation between artifact and text. The need for philological realia knowledge as well as historical ethnoarchaeology suggests that both artifacts and texts have their strong and weak sides; it is as if their "silences" were allocated differently.

The strengths and weaknesses of material culture and writing are due to their different references and hence their different structure. Text is in principle two-dimensional, linear, and built up of a limited number of clearly defined signs, which occur in defined positions. This "limited" character is necessary for all writing, since it is a technologizing of the spoken word. Writing is thus designed to represent the spoken language, and since all oral presentation is linear in both time and space, the text must also preserve this linearity (Ong, 1982). It is precisely this limited character of all texts that has made the decipherment of unknown scripts so similar (Gordon, 1968). Although the meaning of a text can never be unambiguous, text has its most direct access to language as a meaning-given medium, since writing is a representation of the spoken language (Ong, 1982). In contrast, the access of text to the "reality" represented through language is less direct. It may be difficult to visualize the things and the world described in texts, which is why we often need supplementary information in order to interpret a text. This need is the reason why archaeology can sometimes provide philology with details of realia.

Artifacts, in contrast, *are* the world in a wholly different way from texts. The strength of material culture is in fact the very materiality of artifacts, or their form and position in space. Objects are three-dimensional, nonlinear, and composed of an unlimited number of signs that do not appear in fixed positions, since they were not created to render the spoken word. It is rarely possible to decipher artifacts in the same semantic way as texts (see Bloch, 1995; Hodder, 1989; Miller, 1983; Sonesson, 1992). To a much greater extent than writing, material culture has instead both practical and representational functions. Artifacts can be representations, sometimes of such complex ideas as cosmology, which may be traced in many religious monuments, but these ideas cannot be read as a linear text; they rather convey an "expressive force" that can be perceived through sight, hearing, and movement, and that can be described in less strict forms. Although material culture may be full of meaning, this is not expressed as explicitly as in texts, since artifacts do not represent the spoken language. Supplementary information is often required to be able to ascribe meaning to the objects. This quest for explicit meaning from texts is the background to historical ethnoarchaeology. The role of archaeology as philological background knowledge, and the existence of ethnoarchaeology show that there is a reciprocal need for artifacts and texts in broader interpretations of writing as well as material culture. Neither artifact nor text can thus be automatically given primacy of interpretation when these source materials are seen as cultural expressions.

Finally, the relationship between artifacts and texts can be regarded as *discursive contexts*. Neither material culture nor written sources have existed independently of each other or of other forms of human expression. They have been part of complex situations that have also comprised, for example, images, gestures, and oral performances. To understand the discursive contexts, it is important to realize that the relation between the different expressions, and hence the meaning of the different expressions, has shifted in time and place (Moreland, 1991:23). The changing functions of writing are well attested in its 5,000-year history. A very clear example of this is the different uses of cuneiform during the 3,000 years when the script was used (Figure 35; see Larsen 1984). The functions and meanings of images have likewise changed from the first Late Paleolithic cave paintings to today's graffiti. A crucial turning point in the function of pictures and in our way of looking at them came in the latter part of the eighteenth century with the modern concept of art (Belting, 1993). The shifting meaning of

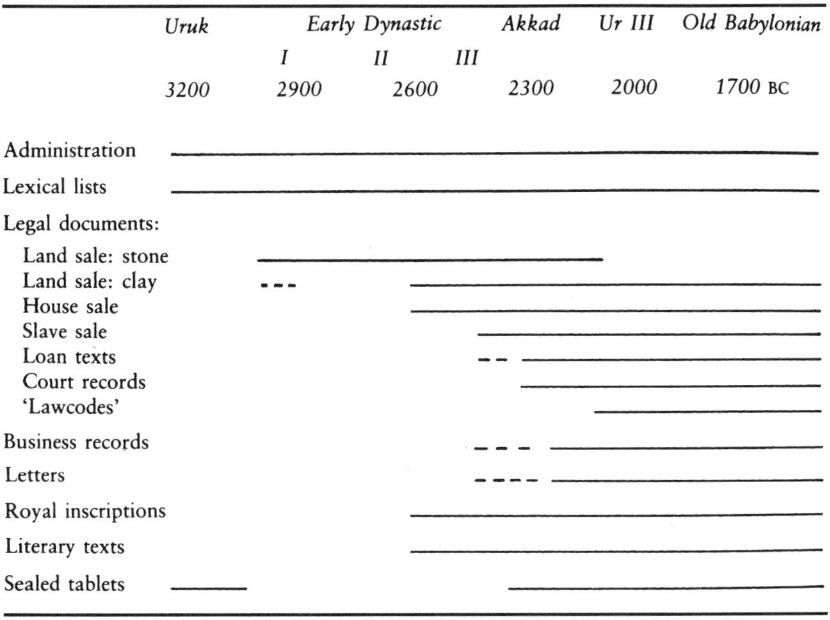

	Uruk	Early Dynastic			Akkad	Ur III	Old Babylonian
		I	*II*	*III*			
	3200	2900	2600		2300	2000	1700 BC

Administration

Lexical lists

Legal documents:
  Land sale: stone
  Land sale: clay
  House sale
  Slave sale
  Loan texts
  Court records
  'Lawcodes'

Business records

Letters

Royal inscriptions

Literary texts

Sealed tablets

Figure 35. The varied applications of cuneiform in Mesopotamia (Postgate, 1992, fig. 3:13, by courtesy of Routledge).

artifacts is also suggested by the way that the traditional chronologies shift from one group of artifacts to another, which shows that the attention to and the meaning of individual categories of artifact have varied through time. John Baines (1988:206) claims that "timeless" pottery, which is of no use for chronologies, has been "emptied" of its symbolic content.

Since the discursive contexts have changed in time and space, the relation between artifact and text is not given. Instead, it must be an important part of work in historical archaeology to try to determine the specific relations. Discursive contexts can be seen historically, for example, through the concepts of oral culture, written culture, and print culture (Little, 1992), but I have instead chosen to make a more analytical presentation, inspired by interart studies (Varga, 1989). From an analytical perspective, the discursive contexts can be described as object-created, integrated, or text-created.

The *object-created* relation can be regarded from both a chronological and a constructivist perspective. The chronological aspect concerns societies that lack writing of their own but that are close in time and place to cultures using texts. The relation thus refers to societies that are primarily dominated by oral tradition and material culture. In oral cultures the tremendous power of the word is almost incredible to people like us who live in a world of texts: Words can kill. The spoken language is thus seen almost as an object. This materiality of the word is typified by the way that oral tradition is often associated with weaving and smithwork (Hampaté Bâ, 1981). People "sew songs" and "forge poems" (Ong, 1982). Yet artifacts are also necessary to give concrete form to oral performances, for instance masks and clothes (see Hampaté Bâ, 1981; Vansina, 1965). The context is to all appearances "prehistoric," but there is an indirect relation to texts, since the object-created relation here refers to societies that are known through contemporary descriptions by outsiders or through later mythological texts. The special thing about the chronological perspective, then, is that it lacks a functional link between artifact and text, but at the same time the relation between material culture and "external" written sources may be important for understanding the society in question.

The constructive aspect of the object-created relation refers instead to the relation between artifact and text in literate societies. The relation can be described as object-created, since the association is expressed through textual commentaries on existing material culture. This association normally consists of direct documentary information about objects, buildings, and places. Most written sources can in fact be linked to this kind of object-created relation. Texts such as contracts of sale, donations, and probate inventories concern material culture, but the artifacts and their production are nevertheless primary to writing.

When the active role of artifacts is emphasized, the constructive aspect also covers other associations. As we have seen, Matthew Johnson (1993:178 ff.) argues that changes in English dwelling houses preceded and partly created Puritanism as a conscious, text-based ideology. The creative role of architecture has similarly been stressed in the history of architecture. Richard Krautheimer (1965:202 ff.) has shown that the final split of Latin Catholicism and Greek orthodoxy in 1054 was not just preceded by a 300-year theological schism but also by a differentiation in ecclesiastical architecture going back 500 years. A special Byzantine style of church architecture had been established as early as the mid-sixth century, with Hagia Sophia in Constantinople. Krautheimer implies that the theological schism was due in part to the fact that liturgy and theology had been adapted to different forms of church architecture for 500 years in the eastern and western Mediterranean.

The second, *integrated*, discursive context consists of situations in which material culture and writing presuppose each other's existence. This applies in the first instance to settings where writing is incorporated in an oral mentality, that is, in most societies before the Late Middle Ages. The fact that writing was created and used in relation to an oral mentality meant that texts in themselves were important objects with a powerful symbolic value. It is therefore important to study how artifact and text have replaced, complemented, and reinforced each other.

Texts as a substitute for artifacts can be illustrated by Clanchy's (1979) study of literacy in medieval England. He is able to show how writing at first had only a secondary role in relation to legally binding, often symbolically charged, acts, which were often performed with special objects. As writing gradually acquired greater legal validity, the importance of ceremonies and artifacts diminished. In other words, it is possible to trace a literal transition from object-based action to text, with the crucial difference being a literate mentality rather than the existence of writing in itself. Postgate (1984) is able to show a comparable change in Babylonia, where symbolic acts declined in significance when cuneiform was applied to a broader range of uses.

Artifact and text may thus presuppose each other. The interpretation of an inscription often depends on the location of the inscription and the design of the object bearing the inscription. As I have said previously, this aspect has been noticed in many contexts, for instance, in studies of Egypt, Greece, China, and Central America. An example comes from boundary stones from ancient Attica in Greece; their inscriptions are so brief that they are incomprehensible unless the position of the boundary stones is preserved or known (Ober, 1995). In other cases, the rhetorical character of artifacts and texts has been used for associations and wordplay. Texts, sculptures, pictures, and monuments have been deliberately created together to mark the significance of a building or a place. The association of the writing with the monument is often evident from the form given to the text itself, since many written languages have a special "epigraphic" form of script, such as the Egyptian hieroglyphs, Latin capitals, and Arabic calligraphy. This emphasis of the monumentality by means of epigraphic script can be seen especially in areas such as Mesopotamia, Egypt, the Mediterranean area, the Arab world, and Central America (see Figure 40).

The third discursive context, which is *text-created*, can be seen from both a constructive and a more communicative perspective. From a constructive point of view, the relation concerns cases where material culture is created directly from writing. The best example of this relation is found when images depict narratives that are known from oral tradition and texts. Formalized iconography of this kind is known, for example, from Egyptian grave decorations, Greek vase paintings, and Central American reliefs. The communicative perspective further stresses that material culture is created after models in image and text. Handwritten pattern books for architects were known in antiquity and the Middle Ages. With the invention of printed books, however, the patterns became of increasing importance. In China printing became a vehicle for thought and education from the tenth century onward (Carter, 1925). In Europe, too, from the second half of the fifteenth century onward, ideas and models began to be mediated through newspapers, pamphlets, pattern books, and guides for all phases of life. Material culture was created not just from gradually acquired skills but more and more with the aid of printed models. The association between print culture and European architecture is fully evident from the Renaissance onward. Through the medium of printed pattern books, architectural styles—including everything from whole buildings to isolated elements such as door surrounds and moldings—could be spread over large areas in a way that had not previously been possible (Figure 36). News of European fashions could similarly be disseminated via pamphlets and newspapers. The printing press has thus made material culture increasingly dependent on printed texts and pictures as mass communication (Eisenstein, 1979).

Artifact and text, then, cannot be separated by a simple stroke of the pen. Material culture and writing may be defined in different ways, depending on whether artifact and text are seen as categories, objects, documents, or discursive contexts. All these perspectives are important, since they all contain possible starting points for further work. I therefore think that we cannot solve the "crisis" of the historical archaeologies solely by stressing one of the perspectives, such as the materiality of text (see Driscoll, 1988; Moreland, 1991). The constant challenge and the constant threat of tautology in the historical archaeologies applies especially to artifact and text as documents and as discursive contexts. Instead of avoiding these perspectives, I think it is important to study them further. In the following section I therefore look more closely at the possibilities and limitations of these perspectives.

THE CONSTRUCTION OF THE CONTEXT

To understand the fundamental methodological dialogue in the historical archaeologies, it is not sufficient to try to define artifact and text. It is also important to study the actual encounter between material

Figure 36. Suggested construction and design of doors in *L'architettura* (1537–1551) by Sebastiano Serlio (1475–1554), here from the German translation (Serlio, 1609(4):66). The picture, which comes from one of the most widely spread Renaissance architectural pattern books, is an example of how printed books, with pictures and commentary, have become increasingly fundamental to material culture.

culture and writing, especially when artifact and text are defined as cultural expressions and discursive contexts. In my view, this encounter can best be described as a special historical archaeological context, created by both artifact and text. By seeing the encounter as a context, which must then be interpreted, one emphasizes the equal status of material culture and writing, unlike traditional perspectives where the text is seen as the given starting point for the interpretation. At the center of the historical archaeological dialogue, then, we have the context and its construction.

All archaeologists are used to working with contexts, for example, the physical context such as stratigraphy, and in recent years the problems and possibilities of context have attracted particular attention in postprocessual or contextual archaeology (see e.g., Hodder and Shanks, 1995). Even though the historical archaeological context is specific, since it is created by both artifact and text, the methodological questions of context can be linked to a more general archaeological debate.

The context is a central but problematic construction in all meaning-producing work. All meaning springs from contexts, and all contexts are in one way or another constructions (Hodder and Shanks, 1995). There are thus no given contexts; their construction is always founded on some preunderstanding or reasonableness associated with different research traditions. What is reasonable in one perspective may be seen as uninteresting or impossible in another. From a methodological point of view, much of the theoretical debate in today's archaeology can be seen as a search for new contexts. For example, reading a rock with images carved on it like a page of a book (Tilley, 1991) puts rock carvings in a completely different context than if the location of the rock is analyzed in relation to the surrounding ecosystem.

In the historical archaeologies it is also clear that the purpose of the context can vary. In the philological and archaeological traditions, the integration of artifact and text is virtually a goal in itself, since it is intended to solve various methodological problems. In the historical and especially the cultural historical traditions, the integration of material culture and writing is rather a means employed to create new and different contexts, which in turn must be interpreted. The new contexts can also be used as hypotheses for new studies (Erdosy, 1988; Schmidt, 1978).

All contexts are built up in a search for similarity and difference. The reasonableness and the preunderstanding in the construction of the context are clearly seen in this search, since "similarity" is not an absolute concept; it is in fact both theoretically and culturally conditioned. Structural similarity, which can be based on concepts such as "symmetry" and "asymmetry," means, for example, that pottery orna-mentation with different forms can be perceived as similar if the pat-terns are symmetrical (Washburn, 1983). Yet even figures that are different can be perceived as having similar forms, such as the circle and the octagon in medieval ecclesiastical architecture. Krautheimer (1942) has pointed out that medieval architects and theologians conceived of the figures as similar since they both expressed a desired "roundness."

The specific historical archaeological context is thus created in a quest for similarity between artifact and text. In view of the vagueness of the concept of similarity, however, the associations between material culture and writing must be seen as analogies (Charlton, 1981). To stress that these analogies are often established with contemporary sources in historical archaeology, I have previously described artifact and text as "contemporary analogies" (Andrén, 1988:18). The question of the con-struction of context in the historical archaeologies can thus be linked to today's growing discussion of analogies in archaeology (see Charlton, 1981; Murray and Walker, 1988; Ravn, 1993; Stahl, 1993; Wylie, 1985). Important features of this debate are the emphasis on the indeter-minability of analogy, as well as the attempts to define the "best" analogies and avoid one-sided analogies.

Closeness in time, space, and form (such as technological sim-ilarity) is always held up as a criterion of a good analogy. The field of historical archaeology can thus be seen as a special case in archaeology as a whole, since the analogies are particularly close because artifact and text are "contemporary analogies." Yet even in the historical archae-ologies, closeness is a relative concept. Text may be written on objects that are important in themselves, and texts can refer to particular monuments and objects, but texts can also concern things that are only indirectly related to material culture, such as ownership in relation to actual traces of cultivation in the landscape (Snodgrass, 1985a). In the latter case, which is common in the historical archaeologies, we lack explicit references to material culture, so it is instead patterns in arti-facts and texts that are compared with one another.

The debate about one-sided analogies links up with the question of whether analogies limit the perspectives in archaeology. Is it possible, for example, to discover wholly unique features in prehistoric societies when archaeological interpretations are of necessity guided by analo-gies from known societies? The methodological solution to this dilemma is to study the past using various analogies and, for every comparison, to look for contradictions between the analogy and the past. By collecting these contrasts or contradictions, it may gradually become possible to arrive at or suggest what is unique (Wylie, 1985). The extensive debate in the historical archaeologies about the dilemma of being dependent on texts can thus be seen as a variant of the discussion about one-sided analogy. In this case too, the solution may be to juxtapose different analogies and to look for contrasts between artifact and text (Leone and Crosby, 1987; Leone and Potter, 1989).

The importance of contrasts in connection with analogy makes it essential to be aware of the differences between the objects under comparison. In this case too, the discussion is relevant for the historical archaeologies and their special context. As will be clear from the previous section about definitions, artifact and text differ as cultural expressions, since their references and structure are different. Their "strength" and their "weakness" are located in different places, which means that their respective weaknesses must be compensated for if we are trying to create coherent contexts of both material culture and writing.

In the following section I present five different types of contexts for the dialogue between artifact and text. They consist of three forms of correspondence, along with association and contrast.

Correspondence

Correspondences between artifact and text can be sought at three levels that are more or less specific: identification, classification, and correlation. Classification normally precedes identification and correlation.

The question of *classification* is fundamental for both prehistoric and historical archaeology (Adams, 1988; Johansen, 1974; Malmer, 1962:37 ff.). But whereas a typology in prehistoric archaeology can and must be constructed without regard to past classifications, it is precisely the relation between past classifications and present-day archaeological definitions that is a fundamental problem in the field of historical archaeology. As we have seen, the same kind of problem is found in anthropology, where the question concerns the relation between the anthropologist's "etic" classification and the "emic" classification of the people under study (see Deagan, 1982; Yentsch, 1989).

Classification has received a great deal of attention in the various historical archaeologies. The heavy dependence on texts in these subjects has meant that classifications known from written sources have long been the starting point for comparisons with material culture. The traditional strategy has been to search deliberately for archaeological "equivalents" to categories known from texts. Finding parallels of this kind is a fundamental feature of iconographic analyses, which build on the fact that the attributes of mythological beings that are known from written sources also appear in pictorial representations. Yet the quest for correspondences may include all types of categories, such as states, cities, different types of monuments such as churches and monasteries, classifications of objects, and the division of people according to gender, age, ethnicity, or social class. A modern example of a text-based classification is Heiko Steuer's (1982, 1989) studies of early medieval graves in Central Europe. He relates differences in burial practice and grave finds directly to the classification of people in early medieval law texts, as this is revealed in the fines paid for manslaughter and murder.

In recent years, however, many scholars have criticized these classifications as a simplified way to harmonize the relation between artifact and text. F. T. Wainwright (1962:107) wrote 35 years ago that "the past is littered with mistaken attempts to equate historical, archaeological and linguistic conceptions." To avoid tendencies to harmonize artifact and text, many scholars have stressed that classifications on the basis of material culture and written documents should first be made independently of each other and only then combined in a search for similarities and differences (e.g., Hsia, 1986; Olsen and Kobyliński, 1991; Samson, 1987). This search should then be seen as a means to create new contexts, which in turn must be interpreted. Classificatory similarity may occur, as in the colonial United States, where pottery designations can be linked to different types of ware (Figure 37). This thereby creates a new context with many possibilities, for example, comparing the pricing of pottery with its occurrence in different settings (Beaudry, 1989). Elsewhere, however, it has proved that the classificatory way of thinking itself differs between past and present. Archaeological typology is normally hierarchically structured, with main types and subtypes, whereas early classification often lacks this hierarchy. In China, jade objects were classified according to Confucian idealism, which means that the designations are very different even when the difference in form is minimal (Hsia, 1986). In other cases the classifications are different because they are based on either function or form. Wheeler showed that Roman pottery was named according to its function, not its form, whereas form is the normal starting point for archaeological typologies (Dark, 1995:86). Similar contradictions have also been found in ethnoarchaeological studies in India (Miller, 1985).

To study more closely whether there is classificatory similarity, it is important to remember that classifications are always constructions. The number of typological elements for archaeological definitions is, as we know, infinite (Malmer, 1962:50 ff.), and designations assigned in the past are similarly one way, out of many conceivable ways, to give

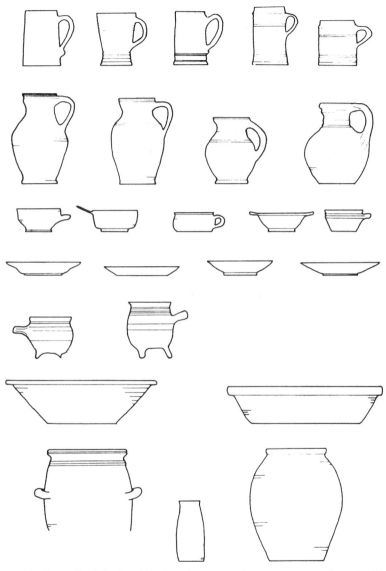

Figure 37. Example of the classification of seventeenth-century pottery from the Chesa-peake region in the present states of Maryland and Virginia, United States (Beaudry et al., 1983:29–37, by courtesy of the Society for Historical Archaeology). The figure shows (from top to bottom) mugs, pitchers, porringers, plates, pipkins, milk pans, and butter pots. The figure is an example of how archaeological typology and designations in old texts can be integrated in a classification that takes account of both form and function.

meaning and structure to the world. This means that categories defined in writing do not necessarily have an immediate physical counterpart. Categories expressed in text may either lack a physical basis or correspond to more or less unambiguous patterns in material culture. An example is the medieval concept of *bod* (related to English *booth*) in Scandinavia. This was a legal term that meant "rented house" or "rented apartment" (Knudsen, 1982). These buildings could vary greatly in appearance, which means that the concept of *bod* does not have any clear material equivalent in the medieval towns.

To establish classificatory similarity, the artifacts and the texts must be made as comparable as possible by compensating for their respective weaknesses as much as we can. Since the strength of artifacts lies in their form and spatial location, most work has been devoted to chronological and functional classification. By dating the artifacts to culturally determined chronologies, one establishes a basic temporal closeness to classifications known from written sources. It is only interesting, for example, to contrast eighteenth-century names for American pottery with excavated pottery when the sherds have also been dated to the eighteenth century. The sharpness of the boundaries for this chronological closeness may vary from case to case. The boundaries are set by periodization of both material culture and linguistic concepts known from texts.

A special feature of many historical archaeologies is that texts are often already present in the data on which dating is based. Inscriptions on buildings, or written records of the foundation of towns and the building of monuments have often played a crucial role in building up chronologies of style. And in many areas, coins are the traditional dating basis for secular settlement and artifact typologies. Radiocarbon datings play a highly subordinate role in the historical archaeologies, since they do not come sufficiently close to the periods known from written sources. Not until the coming of dendrochronology did we obtain a scientific dating method that is independent of texts. Its strength in the historical archaeologies lies not so much in the fact that the method gives "exact" dates, but that it gives an unsurpassed independent closeness to culturally created chronologies that are known from written sources.

Another way to try to make artifacts as comparable with texts as possible is to study their function. Since many categorizations have a functional basis, the function may be an important link between artifact and text. The identification of function may concern anything from individual categories of artifact to entire landscapes. The simplest things are material objects whose form and function are directly associated. Good illustrations of this association come from many religious monuments, which have a strictly designed form linked to their function. This means that little background knowledge is required to classify, say, Christian monasteries, Muslim mosques, Hindu temples, or Buddhist stupas. In many other cases, however, it is much more difficult to determine the function of material culture. Artifacts, rooms, and entire buildings may be multifunctional, thus lacking an unambiguous relation between form and meaning. Deciding whether, for example, a house functioned as a dwelling, a workshop, or a store, or had all three functions at the same time, often requires extensive analyses of the stratigraphic context. Determining the function of material culture is therefore often based on a series of analyses that precede comparisons with written evidence.

With texts as a starting point it is a matter of establishing "closeness" to the artifacts, by giving the content of the text as clear a spatial form as possible. A special form of closeness exists when the writing is found directly on buildings or objects. "Speaking" objects are typical of nascent literacy; they bear reflexive inscriptions such as "I am a comb" or "This is a plane." In some cases, when the function of the object is not obvious, inscriptions of this kind can be important for classification. An example is a special form of painted Maya pottery that has been classified as a vessel for drinking cacao, since the Maya hieroglyph for "cacao" is painted on one example (Stuart, 1988).

Normally, however, it is a matter of taking written descriptions of various defined classes and trying to obtain an idea of their form. An example is the Roman architect Vitruvius's descriptions of ancient architecture. As we saw above, these textual descriptions were insufficient for Alberti, who wanted to create an architecture inspired by antiquity. However, they are sufficiently detailed to enable us to erect parallels with the ancient architecture that is actually known. It is partly thanks to Vitruvius's texts that the different parts of a Roman villa can be named, such as the atrium and the compluvium.

In some cases, an important intermediate stage between artifact and text can be images in a broad sense (see Dymond, 1974:105). They may be murals, mosaic floors, paintings, manuscript illuminations, woodcuts, or book illustrations (see e.g., Groenman-van Waateringe and Velt, 1975; Hasse, 1981). Pictures naturally have a meaning of their own (Gaskell, 1991), but in this context they may play an important mediating role. Through images it is possible to put artifacts into contexts that are difficult to detect archaeologically, and through pictures it is possible to give concepts a physical form that is not evident from written descriptions. An example of the role played by pictures in classifications is Robert Schuyler's (1968) analysis of some indeterminable objects from Fort Louisbourg in Nova Scotia. By comparing the iron objects with some of the 2,900 illustrations in Diderot's *Encyclopédie*, he was able to classify them as ramrods, which meant that they could be used in further analyses of the fort itself.

Besides images, oral tradition has also been compared with material culture. Especially in historical ethnoarchaeology and in historical archaeologies that focus on late periods, such as African archaeology and historical archaeology in the United States, oral tradition plays a major role. It can be used to determine the function of artifact categories and to name them, but it can also be used to trace the meaning of various objects. An oral classification, which does not necessarily agree with an archaeological classification, can sometimes suggest unknown functions and meanings of the objects (see Miller, 1985; Wapnish, 1995).

A fundamental, often implicit, classification normally serves as a basis for the question of *identification*, which is the most specific correspondence in the historical archaeologies. Identification is the hub around which many of the classical problems of the historical archaeologies revolve, especially problems from "early" historical periods, such as Where was Troy? Who was buried at Sutton Hoo? To what does Srivijaya in Sumatra correspond? Identification may concern both time and place. It may concern events such as the foundation of cities, wars, fires, and individual burials, and it may concern places, ranging from geographic areas, via sites, to monuments, individuals, and single objects, such as ancient Greek statues.

Like the discussion of classification, identification problems occupy a large space in the historical archaeologies. This focus on identifications is clearly associated with the heavy dependence on texts in the historical archaeologies. Identification has been a goal in itself, since archaeological remains have been related to political history by means of identified events, persons, and monuments. In addition, the identifications have been used to establish chronologies. The history of Greek art was built up in the nineteenth century by identifying preserved sculptures with sculptures that were known and dated on the basis of ancient descriptions (Bianchi Bandinelli, 1978:49 ff.). Events known from texts, such as city foundations and fires, have likewise been used to date pottery and stratigraphic sequences. In recent decades many scholars have criticized the dominance of identification questions, especially in view of the great risk of arguing in circles (e.g., Barnes, 1984; Chakrabarti, 1984; Franken, 1976; Snodgrass, 1985a). Philip Rahtz (1983) goes so far as to claim that archaeology could just as easily do without the names. It is indeed true to say that it is not interesting just to be able to name an object. An identification becomes interesting only when it can serve as a link between artifact and text and hence be a means to create a new context that is unique to historical archaeology. In other words, identification is interesting as a means, not as an end (Bennett, 1984).

Unlike classification, identity means that a name can normally be expected to have a physical expression, even if the identification cannot be any more than probable (Dymond, 1974:101). The relation may thus seem reasonably simple, but in practice it is often much more complicated. A name may have a diffuse reference so that it is uncertain whether it refers to a place, an area, or merely a "tradition" that is difficult to define. An example is the name Srivijaya in Sumatra, which is traditionally taken to be a kingdom, but may perhaps be the name of a political tradition (Bronson, 1979; see also Mathews, 1991). Towns and villages may have been moved, so that one name refers to several different sites. A monument may have been extended and rebuilt so that a written record does not refer to the entire building but only to one of many stages in its construction. A famous sculpture may have been copied many times, so that a description of it does not fit just the original but also many copies.

As with classification, the method for arriving at identity is based on making artifacts and texts as comparable as possible. It is thus a question of the same type of chronological, spatial, and functional specifications as in classification, intended to compensate for the weak sides of artifacts and texts. Yet since identification concerns more or less unique events and objects, great closeness is often required between artifact and text. Identifying, say, a particular fire layer with a fire at a site known from historical documents often requires very precise datings, since there may be several fire layers close to each other and several fires may have occurred at the site in a short period of time. An optimal form of closeness for identifying objects, buildings, and places is when the artifacts themselves bear texts. This is particularly common in the case of monumental buildings, grave monuments, and sculptures, but it may also occur in connection with portable objects of great symbolic value, such as weapons, jewelry, and drinking vessels.

To make texts comparable with artifacts, it is necessary to create a kind of picture of the past that has a special character. To identify different places in a landscape, one must reconstruct spatial patterns of this landscape via data on such matters as traveling times, distances between places, and their relative position in terms of compass points. These spatial patterns can then be compared with ruined cities and monuments in order to establish a tenable historical topography. This was the technique used by Leake, Robinson, and Cunningham to reconstruct the historical topography of Greece, Palestine, and India (see Figure 13). The same method, although on a smaller scale, is also used to identify sites and buildings within a single place. It is a question of using texts to set up spatial patterns that can then be compared with the actual remains of a city. In this way, for example, palaces, temples, streets, and gates can be identified in a city such as Babylon (Figure 38). Individual objects too, such as Greek sculptures, have been identified by means of specific physical expressions and spatial relations, for example, descriptions of facial expressions and bodily postures (e.g., Niemeyer, 1968). Even if the identification in this case concerns individual objects, it relies on a method by which artifacts and texts are combined through spatial patterns.

Images in a broad sense play an equally important intermediary role in identifications as they do in classifications. Monuments, sculptures, and individuals from ancient Greece, the Roman Empire, medieval Europe, and India can often be identified with the aid of images on coins. In the same way, maps are often a crucial starting point for studies of the historical topography in cities as well as in entire landscapes. In American historical archaeology, early photographs have been used more than once in discussions of identification. An example is the identification of a house in the black settlement of Parting Ways in Plymouth, partly with the aid of photographs (Deetz, 1977:138 ff.).

Oral tradition may be important, as with classification. Information provided by the people who live in a place and use a landscape are often decisive for localizing sites and buildings. Oral tradition can thus be an important element in surveys all over the world (see e.g., Schmidt, 1990). A special form of oral tradition is place-names, which were often used in early studies of historical topography, to detect vanished places known from written sources. In Palestine, Robinson was able to show that the biblical names had survived in Arabic place-names, which thus guided him in his quest for ruins (Moorey, 1991:16).

As with identification, *correlation* presupposes basic classification, but unlike classification and especially identification, correlation is not as much a given relation. There need not be names and concepts with a more or less distinct "counterpart" in material culture. Correlation is instead a search for similar structures or patterns in artifact and text. Correlation is thus concerned more with modern analytical concepts, such as economy; this can be studied by comparing different expressions in artifact and text that can be considered as "economic"; an example would be a comparison of tax rolls with building activity in a particular area (Figure 39).

Unlike classification and identification, correlation is not so much an attempt to compensate for the weak sides of either material culture

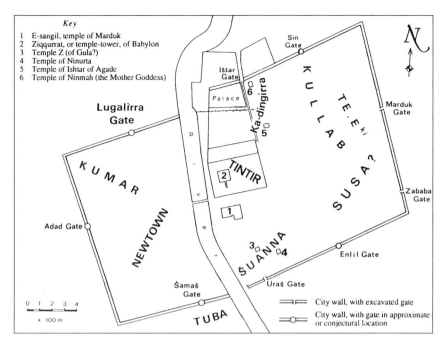

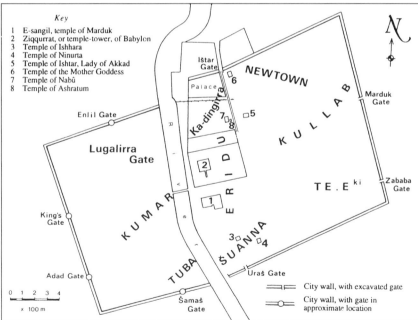

Figure 38. Suggestions from 1930 (top) and 1993 (bottom) for the identification of urban districts and monuments in Babylon (George, 1993, figs 2 and 3, by courtesy of A R George and Antiquity Publications). The pictures are an example of identification as a process, in which new excavations and new texts gradually lead to a reappraisal of the identification of places and buildings.

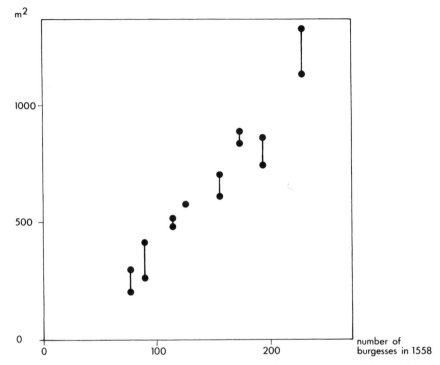

Figure 39. Correlation between the size of town churches 1450–1550 and the number of burghers in 1558 in eight towns in central Denmark (after Andrén, 1985, fig. 19). The size of the town churches is expressed through the area of the interior. The lower value refers to the area of the church itself, whereas the higher value refers to the area of the church, including chapels of more or less private character. The correlation shows that, within a given social and ideological framework, there can be a relatively unambiguous association between monument building and economy. On the other hand, the correlation cannot answer the basic question of why the burghers allocated resources to build churches at all—that problem would require other investigations and other historical archaeological contexts.

or writing. Instead, correlation often assumes that there is an association between the references of artifacts and texts. Correlations are thus based in large measure on perceptions of what is probable, which is ultimately defined by different research traditions. What can be seen in one perspective as an innovative correlation may seem uninteresting or impossible from a different point of view. A typical reflection of the theory-dependence of correlation is that it has above all had an impact in the last few decades, often in more or less explicit attempts to break the tradition of text dependence in the historical archaeologies (see e.g., South, 1977).

Correlation, which may concern time, place, form, and content, often involves attempts to establish correspondence between artifacts and texts that superficially appear not to be connected. The arguments for setting up a correlation may be either quantitative or qualitative. Chronological correlations are built up in the quest for similar periodicity in material culture and writing. This periodicity may be expressed either as chronological breaks or as "cycles" that can be detected in both material culture and writing. An example of chronological correlation is Ian Morris's (1987) study of mortuary practice in the biggest cemetery in Athens, Kerameikos. Morris claims that the chronological patterns he detects in mortuary practice—especially the variability—can be related to different phases in the early political history of Athens. Through correlation, then, mortuary practice is a way for him to detect and to discuss archaeologically the emergence of democracy as a specific form of government for Athens.

Spatial correlations are created in a comparable way, in a quest for parallelism in spatial patterns based on material culture and writing, respectively. These patterns can be both great and small, ranging from individual rooms in a building to whole landscapes. An illustration of a large spatial correlation is Jes Wienberg's (1993) study of late medieval church building in Denmark (see Figure 6). By correlating the spatial variation in the rebuilding and extension of the churches with contemporary tax rolls and soil qualities, he is able to study and discuss the relation between economy and ideology in late medieval Denmark.

Correlations can also be sought without emphasizing the dimensions of time and space. A quantitative example is Stanley South's (1977) demarcation of different artifact patterns. By correlating the frequency of artifacts with different types of settlement, he has been able to define diverse patterns of waste and artifacts from different types of colonial settlements in the United States. A classical example of a qualitative correlation between form and content is the art historian Erwin Panofsky's (1951) interpretation of Gothic cathedrals in France. He argues that the cathedrals were designed in elements and parts in the same way as the theological treatises of scholasticism were composed of chapters and subchapters. The cathedrals and treatises thus reflect a contemporary "mental habit," and according to Panofsky they were two ways of expressing the same thing, namely, Christian faith.

In correlations, as with classifications and identifications, it is a matter of using structural similarities to try to link written evidence with material culture, often as part of a longer argumentation. Correlation thus creates a new context, which must in turn be interpreted. Since it is usually complex patterns that are juxtaposed, total correspondence is never possible; instead only tendencies can be discerned. Since correlation, to a greater extent than classification and identification, is based on probability, more conscious argumentation is often required to create this kind of dialogue between artifact and text.

Association

Association means trying to open an object of study to as many connections as possible. Association occurs in all archaeology and can be both spatial and chronological. Spatial association is particularly significant in stratigraphic analyses and studies of function. It is important, for example, to relate artifacts to buildings in order to understand the function and meaning of a settlement or a monument. In recent years, however, chronological association has also received attention, that is to say, "the past in the past." Several scholars have shown how new buildings or artifacts have been associated with older monuments and incorporated into landscapes with several layers of meaning (Antonacci, 1994; Bradley, 1993; Burström, 1993).

In this connection I consider only one special form of association that is unique to the historical archaeologies, namely, the physical context of writing. This association refers to all the inscriptions and texts discovered by archaeological excavations or inscribed on monuments of various types. The important thing about studying a text and its spatial context is that one creates new contexts that may be fruitful for the interpretation of both material culture and written sources. The strength of the association also lies in the fact that it partly compensates for the weak sides of material culture and writing as cultural expressions. The reference of the writing to "reality" can be made more clear by the find context, while the meaning and function of the artifacts can be made more explicit by the physical presence of the text.

Like correlation, association is not given, however; it is dependent in large measure on research traditions and perspectives. As the survey of the boundary-crossing traditions showed, the role of epigraphy and archaeology in producing texts goes back a long way, but the physical context of writing has not attracted serious attention until the last few decades. This form of association is thus part of the breach with the older text-dominated tradition in the historical archaeologies. The specifically historical archaeological form of association occurs partly in an epigraphic dimension and partly in a documentary dimension. The epigraphic dimension comprises inscriptions on artifacts and monuments that are in themselves important objects, such as temples, painted vases, and jewelry. The inscriptions may be primary, as part of the actual design of the objects, or secondary, as later additions to existing objects, such as graffiti. The primary epigraphic dimension is especially evident in Egypt, Mesopotamia, classical Greece, and Mexico, where the relation between artifact and text can often be perceived as a complement, reinforcement, and variation of a message. Perhaps this association between artifact and text can best be described in the classical terms of rhetoric, such as amplification, hyperbole, and variety (Johannesson, 1990).

An example of primary epigraphic association comes from some Maya temples with stairways consisting of hieroglyphs describing the history of the temple and the city (Figure 40). The texts in these cases underline the monumentality of the temples and their function as a mythological symbol of the place and its political power. The texts could also serve as an additional humiliation for the leaders of defeated enemies, since they could read about the magnificent past of the city on their way up to the top of the temple, where they were sacrificed to the gods (Marcus, 1992:76 ff.). Secondary epigraphic association may be illustrated with the graffiti on the walls of many houses in Pompeii. These inscriptions have been used in several different studies of Pompeii, such as that by Ray Laurence (1994), who has studied "street activity" in the city, partly by comparing the number of graffiti messages on the outer walls of the houses with the length of the streets.

The documentary dimension concerns texts written on objects specially intended for writing, such as cuneiform tablets, Linear B tablets, *mokkan*, birchbark letters, and rune staves. The material on which the text is written is less important here, but the find context of the document is crucial. As we saw from the survey of the subjects above, historical archaeologists have begun to study documentary association in Mesopotamia, Bronze Age Greece, Northern Europe, and Japan (see Figure 10).

The documentary association creates a dialogue between artifact and text that is reciprocally complementary. The texts can provide new perspectives on the function and meaning of the sites, and the find context can simultaneously deepen our understanding of the written documents. One example is Thomas G. Palaima and Cynthia W. Shelmerdine's (1984) analysis of the Mycenaean palace of Pylos. Through the find contexts they are able to interpret one room in the palace as an archive where clay tablets were brought in baskets after the activities described on the tablets were completed. Clay tablets were also found in several small "depots" in other rooms, which were associated with

Figure 40. Maya temple with hieroglyphic stairway at Copán, Honduras (Marcus, 1992, fig. 3.28, by courtesy of Joyce Marcus). The stairway, which contains the longest known Maya text, dated A.D. 750–760, is an illustrative example of how writing and monuments can be integrated with each other in a coherent whole.

ongoing activities. For example, perfume manufacture in the palace can be localized in one room that contained large storage vessels and Linear B tablets about perfume oil.

Another important aspect of the physical context of writing is changes in the inscriptions and their location. In, for example, Mesopotamia, Egypt, and Mexico it was common for old inscriptions to be deliberately destroyed or concealed, as a way to rewrite history. Deposed rulers became nonpersons, and written information about their names and deeds was chiseled away or buried. In other cases, monuments with inscriptions could be carried away by plundering enemies. A well-known example is the stela with the laws of the Babylonian king Hammurabi (1792–1750 B.C.), which was taken as war booty by the Elamites around 1160 B.C. and brought to Susa (Finet, 1973).

In my view, the potential of association lies in the renewal and expansion of the context of both artifacts and texts. However, since association, like other contexts, builds on probability, we must have a clear awareness of the possibilities and limitations of the concepts we use.

Contrast

Contrast is a kind of negative correspondence or a search for differences. The method is found in all archaeology, but in the historical archaeologies it has a special profile, since the contrast can be set up between artifact and text. This specific historical archaeological contrast is normally intended to avoid a one-sided dependence on texts and to problematize written statements on the basis of material culture.

The purpose of contrast, however, can be perceived differently. It can be a way to stress the complexity and the always provisional picture of the past (see Cinthio, 1963). The purpose can also be to trace the "dialectic of history," that is, to detect when material conditions are out of phase with social norms and ideology (see Carmack and Weeks, 1981). Reinhard Wenskus (1979:638) calls for "noncorrespondence," since it helps us to reconstruct an elusive past "reality" and sometimes to define a "historical individuality." Similar ideas are found in Leone, Crosby, and Potter's quest for "ambiguity" between artifact and text, as a way to avoid historical archaeology's dependence on texts and its tautological character (Leone and Crosby, 1987; Leone and Potter, 1989). They start from the discussion of the role of analogy in archaeology. In this debate, ambiguity, difference, anomaly, and refutability (see Binford, 1967; Gould, 1980; Murray and Walker, 1988) have been held up as a methodological way to avoid a total dependence on analogies. By looking for contrasts between the studied objects and the analogies used, it is possible to reach beyond the limits of analogy, to historically unknown things (Wylie, 1985). The quest is thus geared to the same type of "historical individuality" as advocated by Wenskus. Yet the search for contrasts can also be a means to reach people outside the text. Martin Hall (1994) maintains that it is in the differences, or in the "spaces between things and words," that an underclass without access to writing is expressed and can be detected archaeologically.

Contrast is based on probability and research traditions, just like the other contexts. This dependence on perspective is clear in connection with the tricky question of where the difference is located. Since artifact and text, as cultural expressions, differ in character, it is difficult to determine whether discrepancies that are detected are due to shortage of information, lack of comparability, or "actual" differences in the past. There is, for example, no absolute source-critical rule of thumb to decide when there is sufficient information for a difference to be perceived as historically relevant. The dependence on perspective is also seen in the very interest in contrasts. The search for differences between artifact and text is above all a part of the recent breakaway from more text-dominated traditions.

Contrast, like correspondence, between artifact and text can concern time, place, form, or content. It is important, for example, to detect differences in chronological sequences and spatial patterns that can be created with the aid of material culture and writing, respectively. Several deliberate chronological contrasts proceed from the difference between period divisions based on pottery and period divisions based on political and religious history known from texts (Adams, 1979; Chang, 1977; Cinthio, 1963). One of these cases concerns Nubia, where the local pottery chronology cannot be related to political and religious changes, although they comprise, for example, transitions between pharaonic religion, Christianity, and Islam (Adams, 1979). The complex relationship between continuity and change is thus seen in different ways, depending on whether the starting point is texts or artifacts, such as pottery. By combining these perspectives, moreover, a picture of Nubia's history emerges that is unique to it (Figure 41).

In a corresponding way, spatial contrasts have been built up by comparing the occurrence of pottery types with different political territories. Since material culture does not always follow political territories, the picture of different states can be modified (see Keightley, 1983; Morris, 1988; Scott, 1993). An example is Shang-dynasty China, which has been studied by David N. Keightley. He has compared the archaeologically defined "Shang culture" with the political geography of the

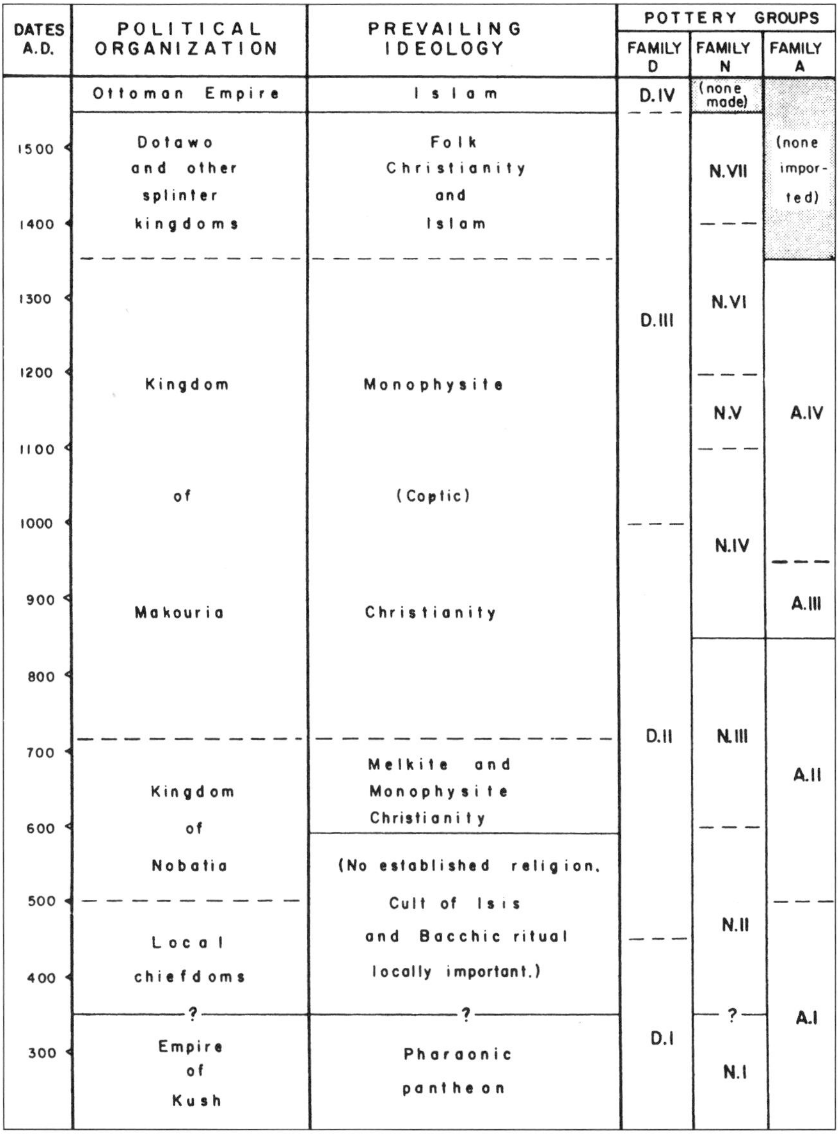

DATES A.D.	POLITICAL ORGANIZATION	PREVAILING IDEOLOGY	POTTERY GROUPS		
			FAMILY D	FAMILY N	FAMILY A
	Ottoman Empire	Islam	D.IV	(none made)	
1500	Dotawo and other splinter	Folk Christianity and		N.VII	(none imported)
1400	kingdoms	Islam		---	
1300			D.III	N.VI	
1200	Kingdom	Monophysite		---	
				N.V	A.IV
1100				---	
1000	of	(Coptic)	---	N.IV	
900	Makouria	Christianity			A.III
800					
700		Melkite and	D.II	N.III	A.II
600	Kingdom of	Monophysite Christianity		---	
	Nobatia	(No established religion.			
500		Cult of Isis		N.II	---
400	Local chiefdoms	and Bacchic ritual locally important.)	---		A.I
	--- ? ---	--- ? ---		--- ? ---	
300	Empire of Kush	Pharaonic pantheon	D.I	N.I	

Figure 41. Comparison of different periodizations of Nubian history (Adams, 1979, fig. 1, by courtesy of The University of Chicago Press). The pottery is divided into local hand-made pottery (D), local wheel-made pottery (N), and imported pottery from Aswan (A). The comparison is an example of chronological contrast between written sources and archaeology.

Shang state, as seen in the 500 or so place-names found on oracle bones
(see Figure 15). The archaeological Shang culture is much larger than
the Shang state of the oracle bones, which Keightley interprets as
showing that the Shang state was part of a larger cultural area and in
itself was equivalent to nothing more than "a thin network of path-
ways and encampments" (Keightley, 1983:523 ff.).

Contrast in form and content can also be sought without any emphasis on the chronological and spatial dimensions. In medieval archaeology and historical archaeology in the United States, several scholars
have studied the difference between, on the one hand, artifact composition in probate inventories and wills and, on the other hand, the artifacts
found archaeologically (e.g., Bowen, 1975; Ideström, 1983; Spencer-
Wood, 1987). Because partly different objects appear in writing and in
archaeological investigations, the differences can be used to reconstruct
a more complete picture of the stock of objects in different settings.
The contrasts can moreover be used to understand how objects were
valued in the past. The things that people left in their wills were not
the same things they threw on their refuse heaps.

Another example of contrast in form and content may be taken from
Martin Hall's (1994) studies of the Cape Colony in the eighteenth century. Descriptions of Cape Town often contain complaints about the
monotonous diet of mutton and the shortage of fish. At the same
time, excavations have shown that there was in fact plenty of fish in
the town. Hall believes that this difference between the material remains and the texts may be connected with the hidden history of the
Cape Colony. Since there were very few European women in the colony,
many of the colonists lived with black slave women. These women
preserved from their original background a strong element of fish in
the diet. Fish was thus associated with slaves, especially female
slaves, in Cape Town, which may explain why fish was less visible than
mutton in European accounts of the place.

A methodologically important aspect of contrast is to study the
actual difference between writing and artifact manufacture. This difference often contains a social dimension, since writing in most societies
has been confined to the elite. In several cases the contrast also contains
a gender dimension, since literacy in many cultures was reserved for
men only (Ong, 1982). Material culture, by contrast, has a more mixed
social and gender background. This social and gender difference between artifact and text is clear, for example, at the Kerameikos cemetery in ancient Athens. Inscriptions and pictures on the gravestones
were, like other "public" monuments, mainly written by and for men (see
Thomas, 1992). The burials, including the deposition of artifacts in

graves, were mainly created by women. Because death above ground was the province of men and death below ground was the sphere of women, it is possible to detect gender differences in the view of the buried people. Above all, the image of woman is divided, since the pictures and texts on the gravestones show the women as chaste housewives and mothers, whereas the grave goods instead arouse associations with the cult of Aphrodite and with women as erotic beings (Houby-Nielsen, 1997). To put it in somewhat extreme terms, the difference between gravestones and grave finds in this context can be described as male texts versus female artifacts.

Contrast is thus an archaeological way to get beyond the text with the aid of texts. Since contrast is a form of negative correspondence, there is no risk of harmonization. Instead, the search for differences requires a distinct awareness of concept formation, of methods, and of the state of our sources as regards both material culture and writing. In addition, the difference between artifact and text is unique to the historical archaeologies, since this negative correspondence is an opportunity to create new and different images of the past.

Smaller and Larger Contexts

I have presented the historical archaeological dialogue as a series of equal contexts built up in a simple relation between artifact and text. In the actual work of relating material culture and writing, however, the connections are often much more complex. As the presentation of the different contexts has shown, several scholars have emphasized that the historical archaeological dialogue must be based on initial independent analyses of artifact and text, to avoid arguing in circles (e.g., Barnes, 1984; Erdosy, 1988). A dilemma in the historical archaeologies is that this methodological stance can be maintained in theory but not always in practice. Many analyses of artifact and text are already subject to the power of scientific tradition from the beginning. As we have seen, datings according to stylistic criteria and artifact chronologies are often based on written sources. And in the same way, many interpretations of words and concepts, when objects have been used as realia knowledge, proceed from material culture. Artifact and text are thus not always mutually independent (see Kosso, 1995).

The contexts have been presented one by one, but in practice they often appear together or as a series of consecutive contexts. Moreover, the texts in the contexts presented here have been depicted as relatively simple and unproblematic statements. In many cases, however, writing can be much more assertive than this. Many early texts are outright historiography, with interpretations of events, places, and monuments. Other texts, such as Ibn Khaldûn's (1332–1406) famous *Muqaddimah* from 1377 (Khaldûn, 1967), represent a philosophy of history with conscious principles for the interpretation of humans and human societies. Other texts are theories of interpretation with instructions for the interpretation of material culture. A well-known example is the four-fold medieval doctrine of interpretation, quadriga, which was a deliberately structured theory of the ambivalence of objects. This interpretative doctrine has successfully been used to interpret, for example, ecclesiastic art from the Middle Ages (e.g., Gotfredsen and Frederiksen, 1988:9 ff.), but the question is whether artifacts are thereby subordinated to texts. Does the doctrine of interpretation mean that the "solution" to problems concerning material culture should nevertheless be sought in texts?

I believe that some form of equality of artifact and text can be maintained even in these cases. Historical writing with interpretations of events is part of the classical area of historical source criticism. Interpretations and perspectives have long been critically examined by historians and should therefore not be accepted outright by archaeologists. Instead, interpretations in older historiography should be seen as statements that can be set alongside material culture in superordinate historical archaeological contexts. In a similar way, more fundamental perspectives, such as philosophy of history and theories of interpretation, may be important parts of a broader context instead of being seen as given answers. The quadripartite medieval doctrine of interpretation may suggest the intention behind a monument or a work of art, but archaeological studies of the same thing may indicate its actual history. In this way, a larger context that contains both theory of interpretation and archaeological investigations can reveal tensions between intention and reality; it is these contrasts that must in turn be interpreted.

The contexts that I have presented here—correspondence, association, and contrast—are different ways of searching for similarity or dissimilarity. On an abstract level, this interplay of similarity and difference is not specific to the historical archaeologies; it is found in all archaeology, as in all meaning-producing work, for instance, in various forms of artistic expression. The same interaction of similarity and dissimilarity is the object of interart studies (Lagerroth et al., 1993). And in the prehistoric archaeologies, classification, correlation, association, and contrast play at least as important a role as in the historical archaeologies. It is just identification that is unique to the historical archaeologies, and—paradoxically—it is scarcely this context that may be expected to lead to a renewal in the historical archaeologies. Instead, the greatest potential for revewal is to be found in correlation, association, and contrast, because new, more or less contradictory contexts can be observed. The unique thing about the historical archaeologies, then, is not the types of context but rather the character of their structure. It is the very dialogue between artifact and text that is unique in relation to prehistoric archaeology as well as history.

Conclusion
Historical Archaeology as a Methodological Perspective

It cannot be taken for granted that subjects such as classical archaeology, Japanese archaeology, Peruvian archaeology, and American historical archaeology should be viewed as a unit. This perspective has been hinted at in the last 25 years (Schuyler, 1970), but it is only in the last 10 years that the view has attracted serious attention. My work is thus one of several contributions to a relatively new archaeological debate that assumes that the boundaries between the individual specialities can and should be crossed. The overall aim is, with the aid of methodological problems, to start a boundary-crossing discussion that will be rewarding both for the individual subjects and for archaeology as a whole.

Even if one recognizes the different historical archaeologies and their diversity, the opinions about the shared problems of method are surprisingly few and similar. The same stances, the same perspectives, and the same methodological solutions recur in subject after subject. It is easy to find discussions in which arguments and stances are "reinvented," or to hear one-sided pleas for a particular perspective, voiced without regard for—or knowledge of—the fact that similar views have been expressed elsewhere. The reason for this parallelism is that the historical archaeologies are part of modern science with its specialization, which is primarily based on the use of different source materials. Even if every subject has its own history and its own research area, all the historical archaeologies share the same dilemma of finding themselves in between more text-based and more artifact-based disciplines. The crucial question is whether archaeology is at all necessary in studies of literate societies. Texts can offer interesting perspectives on and interpretations of material culture, but the presence of texts always runs the risk of making archaeology tautological.

On the basis of this existential question for archaeology I have made a survey of the most important individual subjects as well as of the combined field of historical archaeology. The areas and problems that I have considered are large, but by concentrating on the relation of text and artifact it has been possible to suggest certain perspectives. The historical archaeologies differ in part from prehistoric archaeology, because the sources of inspiration are partly different. In the historical archaeologies the associations with other subjects are expressed through five different methodological approaches that can be seen crossing the boundaries of the individual specialities. The five methodological perspectives also indicate five different ways of looking at artifact and text. The definition of material culture and writing can thereby be perceived as contextual, and the relation between these two human modes of expression as a few constantly recurring contexts.

Since I see historical archaeology as a methodological approach, I have not dealt with the important question of how these contexts can or should be interpreted. As is evident from the survey of the subjects, the five boundary-crossing traditions, and the general historiographical survey, most theoretical perspectives can be applied in the field of historical archaeology. It is thus scarcely possible to recommend certain theoretical perspectives as being particularly suitable for the historical archaeologies. On the other hand, it is possible to detect different strategies that concern not only methodological questions but also empirical and theoretical issues.

A crucial empirical question is whether the historical archaeological dialogue can be conducted by a single person. Is it possible for one and the same person to master artifact and text, and the scientific traditions surrounding them, in such a way that new, interesting, and meaningful contexts really can be constructed? Hildebrand pleaded back in 1882 for an archaeological reading of texts, and the same position is adopted by Beaudry (1989), who calls for an archaeological analysis of texts. Christophersen (1992), in contrast, argues that the special contexts constructed with the aid of both material culture and writing must be created in an interdisciplinary encounter between archaeologists and different textual specialists, since no single person can have equal mastery of the two types of source material. Morris (1994:46) expresses similar fears when he advocates an integrated historical archaeology instead of classical archaeology. The question is who could master all the matter of the new subject that he sketches. The problem of competence probably varies depending on the context. For historical archaeologists in the United States, textual analysis is less of a linguistic problem than it is for, say, classical archaeologists and medieval archaeologists. In certain areas, moreover, archaeology is integrated in area studies that also include language, for example, Egyptology and Assyriology. This integration has been viewed as an important reason for the scant development of archaeology in, for instance, Egypt, but at the same time the assembled study of one area means that the problem of competence is resolved. In my opinion, there is no obvious organizational solution to the problem; instead, it is important to start from existing conditions and try to create as good a dialogue as possible between artifact and text, whether this is done on an individual or a collective basis.

The fundamental methodological problem is whether archaeology is needed in studies of literate societies. To avoid the threat of tautology, clear methodological strategies have been developed in the historical archaeologies. In studies of prehistoric and protohistoric periods, archaeology plays a role that is at once crucial and self-evident. All archaeological results are seen as an asset with the potential to renew our view of the past. The greater the density of texts in a period, however, the more the role of archaeology is questioned, both internally and externally. Archaeological results are scrutinized more critically and may be dismissed as unnecessary confirmations of already known circumstances. To avoid tautological results, five different strategies are used in historical archaeological studies of text-rich periods.

The first strategy emphasizes the methodological role of the historical archaeologies in prehistoric archaeology and to a certain extent philology. Because of their access to texts, the historical archaeologies are perceived as a form of laboratory where archaeological theory and method can be tested and developed. Yet through the link between artifact and text the historical archaeologies can also be seen as a source of knowledge about realia and as a philological reference point. The second, and most common, strategy is to point out the complementary function of archaeology, as a way to study questions or areas of which we know little from written sources, even though there may be many texts. This strategy is a kind of search for more or less text-free zones in time and place where archaeology can be practiced. The third strategy is to proceed from the material dimension of existence, by emphasizing the active role of artifacts in the construction of humans and the human world. The past can be staged in a new material discourse; this was formerly done mainly through historicizing architecture, and today it chiefly takes the form of restorations and museum reconstructions. The past can also be expressed in more text-based historiography, where a correspondence between artifact and text can be accepted, since it does not lead to a simple confirmation of things that are known from written sources. Instead, a correspondence between artifact and text in this cultural historical perspective accentuates the fact that known phenomena can be interpreted in a new way on the basis of a previously unknown material background. A fourth strategy consists in highlighting the materiality of text, which means that writing is analyzed as an archaeological object. By placing more emphasis on the form rather than the content of the text, the assertive character of the text can be disarmed to some extent. At the same time, other aspects of writing and its varying function can be studied, in particular the discursive contexts of text and artifact. The fifth strategy emphasizes the contrast between material culture and writing and hence avoids all simple confirmation of written evidence. Instead the contrast creates a partly new image of the past that is unique to the historical archaeologies.

Finally, theoretical strategies can also be detected in the historical archaeologies. Above all, it is possible to show how the construction of the context is partly theory-bound. All contexts can occur in the various traditions of historical archaeology and in the broad intellectual currents, but the significance of the different contexts clearly varies from one tradition to another. Identification and classification are most important in evolutionary perspectives and can be linked, for example, to philological realia and historical topography. The question of identification is and has been particularly crucial for creating historical identity. By being able to put names on things, places, and areas, both nation-states and colonial powers have been able to claim a form of historical legitimacy, as "heirs" to a magnificent past. On the other hand, correlation and to some extent association and contrast are much more important in the various instances of synchronic and, in a broad sense, functional ideas in the historical archaeologies. Questions about economic and social conditions are studied primarily with the aid of correlation. Finally, it is clear that association and especially contrast are central to today's postmodernism. Association shows the interplay of artifact and text, as well as their rhetorical character, whereas contrast is the difference that gives meaning, as well as showing the indeterminability of the world.

The question is, finally, how the individual historical archaeological disciplines relate to the idea of historical archaeology as a methodological perspective. The subject perspective versus the method perspective exists as a latent opposition throughout this work. Like several other scholars in recent years, I plead for a historical archaeology as a methodological approach, but I simultaneously present the area on the basis of the individual subjects. The field of historical archaeology will probably never be totally united, since ties to specific periods and areas are so strong. Nonetheless, the methodological perspective can have consequences for the individual disciplines (cf. Kardulias, 1994; Morris, 1994), and I shall conclude by illustrating this from my own subject background.

North European medieval archaeology is traditionally demarcated as covering the period of Catholic Christianity in Northern Europe. If the methodological approach is emphasized, then the traditional temporal boundaries of medieval archaeology are of little interest. Post-medieval archaeology is today a fairly new and expanding field of Nordic archaeology. Instead of creating a new discipline, however, this speciality could fit well into an integrated historical archaeology. Likewise, late prehistoric archaeology, which often deals with runic inscriptions, ethnographic descriptions, mythological narratives, place-names, historical analogies, and historical linguistics, could be included in a broader historical archaeology. Instead of a medieval archaeology, as a subject focusing on six centuries of the Nordic past, it would be possible to outline a historical archaeology as a continuous methodological perspective in Northern European archaeology covering at least two thousand years. The early limit for this methodological approach depends on how far ethnographic analogies are accepted (cf. Näsman, 1988; Randsborg, 1993) and whether or not the idea of Indo-European languages and their spread over Europe is accepted and perceived as relevant for archaeology (cf. Kristiansen, 1991; Renfrew, 1987). But regardless of these choices, a coherent historical archaeological perspective distinguishes itself clearly from the subject of medieval archaeology. Historical archaeology as a methodological perspective will no doubt be dominated by direct encounters of artifact and text from the last millennium. However, by underlining the similarities between, say, the debate about language and archaeology and the discussions of artifact and text, the area can also cover aspects of what is normally perceived as prehistoric archaeology. It should be stressed, however, that this applies only to some aspects of prehistoric archaeology, since there are other sides to this field that cannot be considered historical archaeology. Historical archaeology as a methodological perspective might not lead to an archaeology without boundaries, but it may lead to an archaeology with fewer limits.

References

Adams, R. McC., 1966, *The Evolution of Urban Society: Early Mesopotamia and Prehispanic Mexico*. Aldine and Atherton, Chicago and New York.

Adams, R. McC., 1981, *The Heartland of Cities: Surveys of Ancient Settlement and Land Use on the Central Floodplain of the Euphrates*. University of Chicago Press, Chicago.

Adams, R. McC., and Nissen, H. J., 1972, *The Uruk Countryside: The Natural Setting of Urban Societies*. University of Chicago Press, Chicago and London.

Adams, W. Y., 1979, On the Argument from Ceramics to History: A Challenge Based on Evidence from Medieval Nubia. *Current Anthropology* 20:727–734.

Adams, W. Y., 1988, Archaeological Classification: Theory versus Practice. *Antiquity* 61:40–56.

Alberti, L. B., 1988, *On the Art of Building in Ten Books*. English translation of *De re aedificatoria*, by J. Rykwert, N. Leach and R. Tavernor. MIT Press, Cambridge, Massachusetts.

Alkemade, M., 1991, A History of Vendel Period Archaeology: Observations on the Relationship between Written Sources and Archaeological Interpretations. In *Images of the Past: Studies on Ancient Societies in Northwestern Europe*, edited by N. Roymans and F. Theuws, pp. 267–297. Studies in pre- en protohistorie 7. Instituut voor pre- en protohistorische archeologie, Amsterdam.

Allchin, B., and Allchin, F, R., 1982, *The Rise of Civilization in India and Pakistan*. Cambridge University Press, Cambridge.

Allchin, F. R., 1995, Mauryan Architecture and Art. In *The Archaeology of Early Historic South Asia: The Emergence of Cities and States*, edited by F. R. Allchin, pp. 222–273. Cambridge University Press, Cambridge.

An, Z. M., 1989, Chinese Archaeology: Past and Present. *Archaeological Review from Cambridge* 8(1):12–18.

Andah, B. W., 1995, European Encumbrances to the Development of Relevant Theory in African Archaeology. In *Theory in Archaeology: A World Perspective*, edited by P. J. Ucko, pp. 96–109. Routledge, London and New York.

Andersson, H., 1990, *Sjuttiosex medeltidsstäder—aspekter på stadsarkeologi och medeltida urbaniseringsprocess i Sverige och Finland*. Medeltidsstaden 73. Riksantikvarieämbetet och Statens historiska museer, Stockholm.

Andersson, H., and Wienberg, J. (editors), 1993, *The Study of Medieval Archaeology*. Lund Studies in Medieval Archaeology 13. Almqvist & Wiksell International, Stockholm.

Andersson, K., and Forsström, M., 1983, Svensk kyrkoarkeologi. *Hikuin* 9:113–124.

Andrén, A., 1985, *Den urbana scenen: Städer och samhälle i det medeltida Danmark*. Acta Archaeologica Lundensia, Series in 8° Nº 13. Liber, Malmö.

Andrén, A., 1988, Ting och text: Skisser till en historisk arkeologi. *Meta* 1988(1–2):15–28.

Andrén, A., 1994, Medeltidsarkeologi. *Nationalencyclopedin* 13:197. Bra Böcker, Höganäs.

Andrén, A., 1996, Comments on Trade, Towns and States: A Reconsideration of Early Medieval Society. *Norwegian Archaeological Review* 28(2):123–126.

Antonacci, C. M., 1994, Placing the Past: The Bronze Age in the Cultic Topography of

Early Greece. In *Placing the Gods: Sanctuaries and Sacred Space in Ancient Greece*, edited by S. E. Alcock and R. Osborne, pp. 79–104. Clarendon Press, Oxford.

Arnold, C. J., 1986, Archaeology and History: The Shades of Confrontation and Cooperation. In *Archaeology at the Interface: Studies in Archaeology's Relationships with History, Geography, Biology and Physical Science*, edited by J. L. Bintliff and C. F. Gaffney, pp. 32–39. British Archaeological Reports, International Series 300, Oxford.

Ashmore, W., 1992, Deciphering Maya Architectural Plans. In *New Theories on the Ancient Maya*, edited by E. C. Danien and R. J. Sharer, pp. 173–184. The University Museum, University of Philadelphia, Philadelphia.

Astill, G., and Grant, A. (editors), 1988, *The Countryside of Medieval England*. Blackwell, Oxford.

Austin, D., 1990. The 'Proper Study' of Medieval Archaeology. In *From the Baltic to the Black Sea Studies in Medieval Archaeology*, edited by D. Austin and L. Alcock, pp. 9–42. Unwin Hyman, London.

Bagley, R. W., 1992. Changjiang Bronzes and Shang Archaeology. In *Proceedings of the International Colloquium on Chinese Art History 1991, Antiquities Part I*, pp. 214–255. National Palace Museum, Taipei.

Baines, J., 1988, Literacy, Social Organization, and the Archaeological Record: The Case of Early Egypt. In *State and Society: The Emergence and Development of Social Hierarchy and Political Centralization*, edited by J. Gledhill, B. Bender, and M. T. Larsen, pp. 192–214. Unwin Hyman, London.

Ball, J. W., 1986, Campeche, the Itza, and the Postclassic: A Study in Ethnohistorical Archaeology. In *Late Lowland Maya Civilization: Classic to Postclassic*, edited by J. A. Sabloff and E. W. Andrews, pp. 379–408. University of New Mexico Press, Albuquerque.

Bandaranayake, S., 1978, *Arkeologi och imperialism*. Ordfront, Stockholm.

Barker, G., and Lloyd, J. (editors), 1991, *Roman Landscapes: Archaeological Survey in the Mediterranean Region*. Archaeological Monographs of the British School at Rome 2, London.

Barnard, N., 1986, A New Approach to the Study of Clan-Sign Inscriptions of Shang. In *Studies of Shang Archaeology: Selected Papers from the International Conference on Shang Civilization*, edited by K.-C. Chang, pp. 141–206. Yale University Press, New Haven.

Barnes, G. L., 1984, Mimaki and the Matching Game. *Archaeological Review from Cambridge* 3(2):37–47.

Barnes, G. L., 1988, *Protohistoric Yamato: Archaeology of the First Japanese State*. Center for Japanese Studies and the Museum of Anthropology. University of Michigan, Ann Arbor.

Barnes, G. L., 1990, The "Idea of Prehistory" in Japan. *Antiquity* 64:929–940.

Barnes, G. L., 1993, *China, Korea and Japan: The Rise of Civilisation in East Asia*. Thames & Hudson, London.

Barry, T. B., 1987, *The Archaeology of Medieval Ireland*. Methuen, London.

Beaudry, M. C. (editor), 1989, *Documentary Archaeology in the New World*. Cambridge University Press, Cambridge.

Beaudry, M. C., Long, J., Miller, H. M., Neiman, F. D., and Stone, G. W., 1983, A Vessel Typology for Early Chesapeake Ceramics: The Potomac Typological System. *Historical Archaeology* 17(1):18–39.

Beaudry, M. C., Cook, L. J., and Mrozowski, S. A., 1991, Artifacts and Active Voices: Material Culture as Social Discourse. In *The Archaeology of Inequality*, edited by R. H. McGuire and R. Paynter, pp. 150–191. Blackwell, Oxford.

Belting, H., 1993, *Bild und Kult: Eine Geschichte des Bildes vor dem Zeitalter der Kunst*. Beck, Munich.

Bennet, J., 1984, Text and Context: Levels of Approach to the Integration of Archaeological and Textual Data in the Late Bronze Age Aegean. *Archaeological Review from Cambridge* 3(2):63–75.

Bennet, J., 1985, The Structure of the Linear B Administration at Knossos. *American Journal of Archaeology* 89:231–249.

Bent, T., 1892, *The Ruined Cities of Mashonaland*. Longmans, Green, and Co, London.

Bérard, C., Bron, C., Durand, J.-J., Frontisis-Ducroux, F., Lissarrague, F., Schnapp, A., and Vernant, J.-P., 1989, *A City of Images: Iconography and Society in Ancient Greece*. Princeton University Press, Princeton, New Jersey.

Beresford, M., and Hurst, J., 1990, *Wharram Percy: Deserted Medieval Village*. Batsford and English Heritage, London.

Berlo, J. C. (editor), 1983, *Text and Image in Pre-Columbian Art: Essays on the Interrelationship of the Verbal and Visual Arts*. British Archaeological Reports, International Series 180, Oxford.

Bernal, I., 1962, Archaeology and Written Sources. *Akten des 34. Internationalen Amerikanistenkongress, Wien 1960*, pp. 219–225. Ferdinand Berger, Horn and Vienna.

Bernal, I., 1980, *A History of Mexican Archaeology: The Vanished Civilizations of Middle America*. Thames & Hudson, London.

Bernal, M., 1987, *Black Athena: The Afroasiatic Roots of Classical Civilization. Volume 1: The Fabrication of Ancient Greece 1785–1985*. Free Association Books, London.

Bertelsen, R., 1992, En arkeologi for historisk tid eller en europeisk mellomalderarkeologi i Norge? *Meta* 1992(4):9–15.

Bianchi Bandinelli, R., 1978, *Klassische Archäologie: Eine kritische Einführung*. Beck, Munich.

Biddick, K. (editor), 1984, *Archaeological Approaches to Medieval Europe*. XVIII Medieval Institute Publications, Western Michigan University, Kalamazoo.

Bierbrier, M., 1982, *The Tomb-Builders of the Pharaohs*. British Museum Publications, London.

Binford, L. R., 1967. Smudge Pits and Hide Smoking: The Use of Analogy in Archaeological Reasoning. *American Antiquity* 32:1–12.

Binford, L. R., 1977, Historical Archaeology: Is It Historical or Archaeological? In *Historical Archaeology and the Importance of Material Things*, edited by L. Ferguson, pp. 13–22. Society for Historical Archaeology, Tucson.

Bintliff, J. (editor), 1977, *Mycenaean Geography: Proceedings of the Cambridge Colloquium, September 1976*. British Association for Mycenaean Studies, Cambridge.

Bjørneboe, L., 1975, *Inkarigets samfund*. Gyldendal, Copenhagen.

Bleed, P., 1986, Almost Archaeology: Early Archaeological Interest in Japan. In *Windows on the Japanese Past: Studies in Archaeology and Prehistory*, edited by R. J. Pearson, pp. 57–67. Center for Japanese Studies, University of Michigan, Ann Arbor.

Bloch, M., 1995, Questions Not to Ask the Malagasy Carvings. In *Interpreting Archaeology: Finding Meaning in the Past*, edited by I. Hodder, M. Shanks, A. Alexandri, V. Buchli, J. Carman, J. Last, and G. Lucas, pp. 212–215. Routledge, London and New York.

Blomqvist, R., 1941, *Tusentalets Lund*. Skrifter utgivna av föreningen Det Gamla Lund 21–22, Lund.

Bowen, J., 1975, Probate Inventories: An Evaluation from the Perspective of Zooarchaeology and Agricultural History at Mott Farm. *Historical Archaeology* 9:11–25.

Bradley, R., 1993, *Altering the Earth: The Origins of Monuments in Britain and Continental Europe*. Society of Antiquaries of Scotland, Edinburgh.

Brandt, S. A., and Fattovich, R., 1990, Late Quaternary Archaeological Research in the Horn of Africa, In *A History of African Archaeology*, edited by P. Robertshaw, pp. 95–108. James Currey, London.

Bratton, F. G., 1967, *A History of Egyptian Archaeology*. Robert Hale, London.

Braunfels, W., and Schnitzler, H. (editors), 1965, *Karolingische Kunst*. Karl der Grosse. Lebenswerk und Nachleben 3. L. Schwann, Düsseldorf.

Bredsdorff, T., Larsen, M., and Thyssen, O., 1979, *Til glæden: Om humanisme og humaniora*. Gyldendal, Copenhagen.

Bronson, B., 1979, The Archaeology of Sumatra and the Problem of Srivijaya. In *Early South East Asia: Essays in Archaeology, History and Historical Geography*, edited by R. B. Smith and W. Watson, pp. 315–341. Oxford University Press, New York and Oxford.

Brown, M. R., 1989, The Behavioral Context of Probate Inventories: An Example from Plymouth Colony. In *Documentary Archaeology in the New World*, edited by M. C. Beaudry, pp. 79–82. Cambridge University Press, Cambridge.

Bruneau, P., 1974, Sources textuelles et vestiges matériels: Réflexions sur l'interprétation archéologique. In *Mélanges helléniques offerts à Georges Daux*, pp. 33–42. Editions E. De Boccard, Paris.

Budde, H., 1993, Japanische Farbholzschnitte und europäische Kunst: Maler und Sammler im 19. Jahrhundert. In *Japan und Europa 1543–1929*, edited by D. Croissant and L. Ledderose, pp. 164–177. Argon, Berlin.

Bureus, J., 1664, *Monumenta Lapidum aliquot Rvnicorvm*. Curio, Uppsala.

Burger, R. L., 1989, An overview of Peruvian archaeology (1976–1986). *Annual Review of Anthropology* 18:37–69.

Burke, P. (editor), 1991, *New Perspectives on Historical Writing*. Polity Press, Oxford.

Burström, M., 1993, *Mångtydiga fornlämningar: En studie av innebörder som tillskrivits fasta fornlämningar i Österrekarene härad, Södermanland*. Stockholm Archaeological Reports 27, Stockholm.

Capon, E., 1977, *Art and Archaeology in China*. Macmillan, South Melbourne.

Carmack, R., and Weeks, J., 1981, The Archaeology and Ethnohistory of Utatlan: A Conjunctive Approach. *American Antiquity* 46:323–341.

Carter, E. F., and Stolper, M. W., 1984, *Elam: Surveys of Political History and aArchaeology*. University of California, Publications in Near Eastern Studies 25. Berkeley.

Carter, T. F., 1925, *The Invention of Printing in China and Its Spread Westward*. Columbia University Press, New York.

Chakrabarti, D. K., 1984, Archaeology and the Literary Tradition: An Examination of the Indian Context. *Archaeological Review from Cambridge* 3(2):29–36.

Chakrabarti, D. K., 1988. *A History of Indian Archaeology: From the Beginning to 1947*. Munishiram Manoharlal, New Delhi.

Champion, T. C., 1985, Written Sources and the Study of the European Iron Age. In *Settlement and Society: Aspects of West European Prehistory in the First Millennium BC*, edited by T. C. Champion and J. V. S. Megaw, pp. 9–22. Leicester University Press, Leicester.

Champion, T. C., 1990, Medieval Archaeology and the Tyranny of the Historical Record. In *From the Baltic to the Black Sea: Studies in Medieval Archaeology*, edited by D. Austin and L. Alcock, pp. 79–95. Unwin Hyman, London.

Chang, K.-C., 1976, *Early Chinese Civilization: Anthropological Perspectives*. Harvard University Press, Cambridge, Massachusetts.

Chang, K.-C., 1977, *The Archaeology of Ancient China*. Yale University Press, New Haven.

Chang, K.-C., 1981. Archaeology and Chinese Historiography. *World Archaeology* 13: 156–169.

Chang, K.-C., 1983, *Art, Myth and Ritual: The Path to Political Authority in Ancient China*. Harvard University Press, Cambridge, Massachusetts.

Chang, K.-C., 1986, *The Archaeology of Ancient China*. Yale University Press, New Haven.

Chang, K.-C., 1989, Ancient China and Its Anthropalogical Significance. In *Archaeological Thought in America*, edited by C. C. Lamberg-Karlovsky, pp. 155–166. Cambridge University Press, Cambridge.

Chapelot, J., and Fossier, R. (editors), 1985, *The Village and House in the Middle Ages*. Batsford, London.

Charlton, T. H., 1981, Archaeology, Ethnohistory, and Ethnology: Interpretive Interfaces. In *Advances in Archaeological Method and Theory*, vol. 4, edited by M. B. Schiffer, pp. 129–176. Academic Press, New York.

Charnay, D., 1887, *The Ancient Cities of the New World*. Chapman and Hall, London.

Chávez, S. J., 1992, A Methodology for Studying the History of Archaeology: An Example from Peru (1524–1900). In *Rediscovering Our Past: Essays on the History of American Archaeology*, edited by J. E. Reyman, pp. 35–50. Avebury, Aldershot.

Chaze, D. Z., 1986, Social and Political Organization in the Land of Cacao and Honey: Correlating the Archaeology and Ethnohistory of the Postclassic Lowland Maya. In *Late Lowland Maya Civilization: Classic to Postclassic*, edited by J. A. Sabloff and E. W. Andrews V, pp. 347–378. University of New Mexico Press, Albuquerque.

Chen, C., 1989, Chinese Archaeology and the West. *Archaeological Review from Cambridge* 8(1):27–35.

Chippendale, C., 1991, Editorial. *Antiquity* 65:439–446.

Chrétien, J.-P., 1986, Confronting the Unequal Exchange of the Oral and the Written. In *African Historiographies: What History for Which Africa?* edited by B. Jewsiewicki and D. Newbury, pp. 75–90. Sage Publications, Beverly Hills, London, and New Delhi.

Christophersen, A., 1979, Arkeologi: Mer enn et hull i jorden? Synspunkter på forholdet mellom arkeologi og historie. *Meta* 1979 (1):4–8; (2):4–8.

Christophersen, A., 1980, *Håndverket i forandring. Studier i horn- og beinhåndverkets utvikling i Lund ca. 1000–1350*. Acta Archaeologica Lundensia, Series in 4° N° 13. Gleerups, Lund.

Christophersen, A., 1992, Mellom tingenes tale og tekstenes tyranni: Om faglig identitet og selvforståelse i historisk arkeologi. *Meta* 1992(4):70–87.

Cinthio, E., 1957, *Lunds domkyrka under romansk tid*. Acta Archaeologica Lundensia, Series in 8° N° 1. Rudolf Habelt, Bonn, and Gleerups, Lund.

Cinthio, E., 1963, Medieval Archaeology as a Research Subject. *Meddelanden från Lunds universitets historiska museum* 1962–1963: 186–202.

Cinthio, E., 1968, Kyrkorummet—funktion och utsmyckning. In *Skånsk lantkyrka från medeltiden*, by E. Gustafsson and E. Cinthio, pp. 67–133. Sydsvenska Dagbladets årsbok 1969, Malmö.

Cinthio, E., 1984, Vad betyder medeltiden för arkeologin? In *Den historiska tidens arkeologi i Finland*, edited by H. Brusila, pp. 52–64. Åbo landskapsmuseums rapporter 6, Åbo.

Clanchy, M. T., 1979, *From Memory to Written Record: England 1066–1307*. Edward Arnold, London.

Clarke, D., 1971, Archaeology: The Loss of Innocence. *Antiquity* 47:6–18.

Clarke, H., 1984, *The Archaeology of Medieval England*. British Museum Publications, London.

Coe, M. D., 1992, *Breaking the Maya Code*. Thames & Hudson, London.

Collett, D. P., 1993, Metaphors and Representations Associated with Precolonial Iron Smelting in Eastern and Southern Africa. In *The Archaeology of Africa: Food, Metals and Towns*, edited by T. Shaw, P. J. J. Sinclair, B. Andah, and A. Okpoko, pp. 499–511. Routledge, London and New York.

Colley, S., 1995, What Happened at WAC-3? *Antiquity* 69:15–18.

Collins, J., 1995, Literacy and Literacies. *Annual Review of Anthropology* 24:75–93.

Connah, G., 1987, *African Civilization: Precolonial Cities and States in Tropical Africa: An Archaeological Perspective*. Cambridge University Press, Cambridge.

Connah, G., 1988, *"Of the Hut I Builded": The Archaeology of Australia's History*. Cambridge University Press, Cambridge.

Conningham, R. A. E., 1995, Dark Age or Continuum? An Archaeological Analysis of the Second Emergence of Urbanism in South Asia. In *The Archaeology of Early Historic South Asia: The Emergence of Cities and States*, edited by F. R. Allchin, pp. 54–72. Cambridge University Press, Cambridge.

Cook, R. M., 1972, *Greek Painted Pottery*. Methuen, London.

Copley, G., 1986, *Archaeology and Place-Names in the Fifth and Sixth Centuries*. British Archaeological Reports, British Series 147, Oxford.

Coquery-Vidrovitch, C., and Jewsiewicki, B., 1986, Africanist Historiography in France and Belgium: Traditions and Trends. In *African Historiographies: What History for Which Africa?* edited by B. Jewsiewicki and D. Newbury, pp. 139–150. Sage Publications, Beverly Hills, London, and New Delhi.

Cordy-Collins, A., 1983, Ancient Andean Art as Explained by Andean Ethnohistory: An Historical Review. In *Text and Image in Pre-Columbian Art: Essays an the Interrelationship of the Verbal and Visual Arts*, edited by J. C. Berlo, pp. 181–196. British Archaeological Reports, International Series 180, Oxford.

Coulmans, F., 1996, *The Blackwell Encyclopedia of Writing Systems*. Blackwell, Oxford.

Croissant, D., and Ledderose, L. (editors), 1993, *Japan und Europa 1543–1929*. Argon, Berlin.

Culbert, T. P. (editor), 1991, *Classic Maya Political History: Hieroglyphic and Archaeological Evidence*. Cambridge University Press, Cambridge.

Cunningham, A., 1871, *Archæological Survey of India I: Four Reports Made during the Years 1862–63–64–65*. The Government Central Press, Simla.

Curl, J. S., 1982. *The Egyptian Revival: An Introductory Study of a Recurring Theme in the History of Taste*. George Allen & Unwin, London.

Daggett, R. E., 1992, Tello, The Press and Peruvian Archaeology. In *Rediscovering Our Past: Essays on the History of American Archaeology*, edited by J. E. Reyman, pp. 191–202. Avebury, Aldershot.

d'Agostino, B., 1991, The Italian Perspective on Theoretical Archaeology. In *Archaeological Theory in Europe: The Last Three Decades*, edited by I. Hodder, pp. 52–64. Routledge, London and New York.

D'Altroy, T. N., and Hastorf, C. A., 1992, The Architecture and Contents of Inka State Storehouses in the Xanxa Region of Peru. In *Inka Storage Systems*, edited by T. Y. LeVine, pp. 259–286. University of Oklahoma Press, Norman and London.

Daniel, G., 1981, *A Short History of Archaeology*. Thames and Hudson, London.

Dark, K. R., 1995, *Theoretical Archaeology*. Duckworth, London.

Darling, P. J., 1984, *Archaeology and History in Southern Nigeria: The Ancient Linear Earthworks of Benin and Ishan*. British Archaeological Reports, International Series 215, Oxford.

Davies, P., 1985, *Splendours of the Raj: British Architecture in India 1660 to 1947*. John Murray, London.

Dawson, R., 1967, *The Chinese Chameleon: An Analysis of European Conceptions of Chinese Civilization*. Oxford University Press, London.

Deagan, K., 1982, Avenues of Inquiry in Historical Archaeology. In *Advances in Archaeological Method and Theory*, vol. 5, edited by M. B. Schiffer, pp. 151–177. Academic Press, New York.

Deagan, K., 1983, *Spanish St. Augustine: The Archaeology of a Colonial Creole Community*. Academic Press, New York.

de Barros, P., 1990, Changing Paradigms, Goals & Methods in the Archaeology of Francophone West Africa. In *A History of African Archaeology*, edited by P. Robertshaw, pp. 155–172. James Curry, London.

de Bouard, M., 1969, The Centre for Medieval Archaeological Research, University of Caen. *World Archaeology* 1:61–67.

de Bouard, M., 1975, *Manuel d'archéologie médiévale: De la fouille à l'histoire*. S.E.D.E.S., Paris.

DeCorse, C. R., 1993, The Danes on the Gold Coast: Culture Change and the European Presence. *African Archaeological Review* 11:149–173.

Deetz, J., 1977, *In Small Things Forgotten: The Archaeology of Early American Life*. Anchor Books/Doubleday, Garden City, New York.

Deetz, J., 1991, Introduction: Archaeological Evidence of Sixteenth- and Seventeenth-Century Encounters. In *Historical Archaeology in Global Perspective*, edited by L. Falk, pp. 1–9. Smithsonian Institution Press, Washington D.C. and London.

Deetz, J., and Dethlefsen, E., 1965, The Doppler Effect and Archaeology: A Consideration of the Spatial Aspects of Seriation. *Southwestern Journal of Anthropology* 21:196–206.

Deetz, J., and Dethlefsen, E., 1967, Death's Head, Cherub, Urn and Willow. *Natural History* 76(3):29–37.

de Maret, P., 1990, Phases & Facies in the Archaeology of Central Africa. In *A History of African Archaeology*, edited by P. Robertshaw, pp. 109–134. James Currey, London.

Denon, V., 1802, *Voyage dans la basse et haute Égypte, pendant les campagnes du général Bonaparte*. Didot L'Aîné, Paris.

Deshpande, M. N., 1967, Historical Archaeology. In *Review of Indological Research in the Last 75 Years*, edited by P. J. Chimmulgund and V. V. Mirashi, pp. 419–458. Vinayak S. Chitrao, Poona.

Dethlefsen, E., and Deetz, J., 1966, Death's Heads, Cherubs, and Willow Trees: Experimental Archaeology in Colonial Cemeteries. *American Antiquity* 31:502–510.

Dever, W. G., 1972, *Archaeology and Biblical Studies: Retrospects and Prospects*. Seabury-Western Theological Seminary, Evanston.

Dever, W. G., 1990, *Recent Archaeological Discoveries and Biblical Research*. University of Washington Press, Seattle and London.

Dickens, R. S., Jr. (editor), *Archaeology of Urban America: The Search for Pattern and Process*. Academic Press, New York.

Diringer, D., 1962, *Writing*. Ancient Peoples and Places 25. Thames & Hudson, London.

Diringer, D., 1968, *The Alphabet: A Key to the History of Mankind I–II*. Hutchinson, London.

Djait, H., 1981, Written Sources before the Fifteenth Century. In *Unesco General History of Africa, I. Methodology and African Prehistory*, edited by J. Ki-Zerbo, pp. 87–113. Unesco, Paris and London.

Dohme, R., 1887, *Geschichte der Deutschen Baukunst*. Grote, Berlin.

Donley, L. W., 1982, House Power: Swahili Space and Symbolic Markers. In *Symbolic and Structural Archaeology*, edited by I. Hodder, pp. 63–73. Cambridge University Press, Cambridge.

Driscoll, S. T., 1988, The Relationship between History and Archaeology: Artefacts, Documents and Power. In *Power and Politics in Early Medieval Britain and Ireland*, edited by S. T. Driscoll and M. R. Nieke, pp. 162–187. Edinburgh University Press, Edinburgh.

Dymond, D. P., 1974, *Archaeology and History: A Plea for Reconciliation*. Thames & Hudson, London.

Dyson, S. L., 1993, From New to New Age Archaeology: Archaeological Theory and Classical Archaeology—A 1990s Perspective. *American Journal of Archaeology* 97: 195–206.

Dyson, S. L., 1995, Is There a Text in This Site? In *Methods in the Mediterranean. Historical and Archaeological Views on Texts and Archaeology*, edited by D. B. Small, pp. 25–44. Mnemosyne, Supplement 135. E. J. Brill, Leiden, New York, and Cologne.

Earle, T. K., and D'Altroy, T. N., 1989, The Political Economy of the Inka Empire: The Archaeology of Power and Finance. In *Archaeological Thought in America*, edited by C. C. Lamberg-Karlovsky, pp. 183–204. Cambridge University Press, Cambridge.

Ehret, C., and Posnansky, M. (editors), 1982, *The Archaeological and Linguistic Reconstruction of African History*. University of California Press, Berkeley.

Eisenstein, E., 1979, *The Printing Press as an Agent of Change: Communications and Cultural Transformations in Early-Modern Europe* 1–2. Cambridge University Press, Cambridge.

Ellis, M. de Jong, 1983, Correlation of Archaeological and Written Evidence for the Study of Mespotamian Institutions and Chronology. *American Journal of Archaeology* 87:497–507.

Emt, J., and Hermerén, G. (editors), 1990, *Konst och filosofi: Texter i estetik*. Studentlitteratur, Lund.

Erdosy, G., 1988, *Urbanisation in Early Historic India*. British Archaeological Reports, International Series 430, Oxford.

Fabian, J., 1983, *Time and the Other: How Anthropology Makes its Object*. Columbia University Press, New York.

Farrington, I. S., 1992, Ritual Geography, Settlement Patterns and the Characterization of the Provinces of the Inka Heartland. *World Archaeology* 23(3):368–385.

Fash, W. L., 1988, A New Look at Maya Statecraft from Copán, Honduras. *Antiquity* 62:157–169.

Fash, W, L., 1994, Changing Perspectives on Maya Civilization. *Annual Review of Anthropology* 23:181–208.

Fehring, G. P., 1991, *The Archaeology of Medieval Germany: An Introduction*. Routledge, London.

Felgenhauer-Schmiedt, S., 1993, *Die Sachkultur des Mittelalters im Lichte der archäologischen Funde*. Peter Lang, Frankfurt am Main.

Fine, J. V. A., 1983, *The Early Medieval Balkans: A Critical Survey from the Sixth to the Late Twelfth Century*. The University of Michigan Press, Ann Arbor.

Fine, J. V. A., 1987, *The Late Medieval Balkans: A Critical Survey from the Late Twelfth Century to the Ottoman Conquest*. The University of Michigan Press, Ann Arbor.

Finet, A., 1973, *Le Code Hammurabi*. Les éditions du Cerf, Paris.

Finley, M. I., 1971, Archaeology and History. *Daedalus* 100(1):168–186.

Flannery, K. V., 1982, *Maya Subsistence*. Academic Press, New York.

Flannery, K. V., and Marcus, J. (editors), 1983, *The Cloud People: Divergent Evolution of the Zapotec and Mixtec Civilizations*. Academic Press, New York.

Fletcher, R., 1993, Settlement Area and Communication in African Towns and Cities. In *The Archaeology of Africa: Food, Metals and Towns*, edited by T. Shaw, P. J. J. Sinclair, B. Andah, and A. Okpoko, pp. 732–747. Routledge, London and New York.

Fong, W. (editor), 1980, *The Great Bronze Age of China*. The Metropolitan Museum of Art, New York.

France, P., 1991, *The Rape of Egypt: How the Europeans Stripped Egypt of Its Heritage*. Barrie & Jenkins, London.

Francovich, R., 1989, The Makings of Medieval Tuscany. In *The Birth of Europe: Archaeology and Social Development in the First Millenium* A.D., edited by K. Randsborg, pp. 166–172. Analecta Romana Instituti Danici, Supplementum XVI. L'Erma de Bretschneider, Rome.

Francovich, R., 1993, Some Notes on Medieval Archaeology in Mediterranean Europe. In *The Study of Medieval Archaeology*, edited by H. Andersson and J. Wienberg, pp. 49–62. Lund Studies in Medieval Archaeology 13. Almqvist & Wiksell International, Stockholm.

Francovich, R., and Parenti, R. (editors), 1988, *Archaeologia e restauro dei monumenti*. Giglio, Florence.

Frandsen, B. S., 1993, Fascismen og det romerske imperium. In *Imperium Romanum: Realitet, idé, ideal III*, edited by S. O. Due and J. Isager, pp. 97–123. Sfinx, Århus.

Franken, H. J., 1976, The Problem of Identification in Biblical Archaeology. *Palestine Exploration Quarterly* 108:3–11.

Fraser, D., and Cole, H. M. (editors), 1972, *African Art and Leadership*. University of Wisconsin Press, Madison.

Friedman, J., 1994, Modernitetens implosion. In *Kulturen i den globala byn*, edited by O. Hemer, pp. 17–36. Ægis, Lund.

Fritz, J. M., and Mitchell, G., 1987, Interpreting the Plan of a Medieval Hindu Capital, Vijayanagara. *World Archaeology* 19:105–129.

Funari, P. P. A., 1995, Mixed Features of Archaeological Theory in Brazil. In *Theory in Archaeology: A World Perspective*, edited by P. J. Ucko, pp. 236–250. Routledge, London and New York.

Gaimster, D., 1994, The Archaeology of Post-Medieval Society, c.1450–1750: Material Culture Studies in Britain Since the War. In *Building on the Past: Papers Celebrating 150 Years of the Royal Archaeological Institute*, edited by B. Vyner, pp. 283–312. Royal Archaeological Institute, London.

Gaskell, I., 1991, History of Images. In *New Perspectives on Historical Writing*, edited by P. Burke, pp. 168–192. Polity Press, Oxford.

George, A. R., 1993, Babylon Revisited: Archaeology and Philology in Harness. *Antiquity* 67:734–746.

Ghaidan, U. (editor), 1976, *Lamu: A Study of Conservation*. East African Literature Bureau, Nairobi, Kampala, and Dar-es-Salaam.

Gilchrist, R., 1994, *Gender and Material Culture: The Archaeology of Religious Women*. Routledge, London and New York.

Goody, J., 1986, *The Logic of Writing and the Organization of Society*. Cambridge University Press, Cambridge.

Gordon, C. H., 1968, *Forgotten Scripts: How They Were Deciphered and Their Impact on Contemporary Culture*. Basic Books, New York.

Gotfredsen, L., and Frederiksen, H. J., 1988, *Troens billeder: Romansk kunst i Danmark*. Systime, Herning.

Gould, R. A., 1980, *Living Archaeology*. Cambridge University Press, Cambridge.

Gould, R. A. (editor), 1978, *Explorations in Ethnoarchaeology*. University of New Mexico Press, Albuquerque.

Gould, S. J., 1981, *The Mismeasure of Man*. Norton, New York.

Grabar, O., 1971, Islamic Archaeology: An Introduction. *Archaeology* 24(3):197–199.

Grandien, B., 1974, *Drömmen om medeltiden: Carl Georg Brunius som byggmästare och idéförmedlare*. Nordiska Museets handlingar 82, Stockholm.

Greene, K., 1986, *The Archaeology of the Roman Economy*. Batsford, London.

Groenman-van Waateringe, W., and Velt, L. M., 1975, Schulmode im späten Mittelalter: Funde und Abbildungen. *Zeitschrift für Archäologie des Mittelalters* 3:95–119.

Halbfass, W., 1981, *Indien und Europa: Perspektiven ihrer geistigen Begegnung.* Schwabe, Basle and Stuttgart.

Hall, M., 1990, "Hidden History," Iron Age Archaeology in Southern Africa. In *A History of African Archaeology*, edited by P. Robertshaw, pp. 59–77. James Currey, London.

Hall, M., 1993, The Archaeology of Colonial Settlement in Southern Africa. *Annual Review of Anthropology* 22:177–200.

Hall, M., 1994, Subaltern Voices? Finding the Spaces between Things and Words. Paper presented at the World Archaeological Congress 3, New Delhi, December 1994.

Hammond, N., 1983, Lords of the Jungle: A Prosopography of Maya Archaeology. In *Civilization in the Ancient Americas*, edited by R. M. Leventhal and A. L. Kolata, pp. 3–32. University of New Mexico Press, Albuquerque.

Hampaté Bâ, A., 1981, The Living Tradition. In *Unesco General History of Africa, I. Methodology and African Prehistory*, edited by J. Ki-Zerbo, pp. 166–205. Unesco, Paris and London.

Hardesty, D. C., 1988, *The Archaeology of Mining and Miners: A View from the Silver State.* Society for Historical Archaeology, Tucson.

Harnow, H., 1992, Industriel arkæologi—modefænomen eller tiltrængt nybrud? *Fortid og Nutid* 39:253–271.

Harrington, J. C., 1955, Archaeology as Auxiliary to American History. *American Anthropologist* 57:1121–1130.

Harrington, J. C., 1957, *New Light on Washington's Fort Necessity.* Eastern National Park and Monument Association, Richmond, Virginia.

Harris, M., 1968, *The Rise of Anthropological Theory.* Crowell, New York.

Harris, R., 1986, *The Language Myth.* Duckworth, London.

Harvey, D., 1989, *The Condition of Post-Modernity.* Blackwell, Oxford.

Haskell, F., 1993, *History and Its Images: Art and the Interpretation of the Past.* Yale University Press, New Haven and London.

Hasse, M., 1981, Neues Hausgerät, Neue Häuser, Neue Kleider—Eine Betrachtung der städtischen Kultur im 13. und 14. Jahrhundert, sowie ein Katalog der metallenen Hausgerät. *Zeitschrift für Archäologie des Mittelalters* 7:7–83.

Hautala, J., 1971, Folkminnesforskningen som vetenskapsgren. In *Folkdikt och folktro*, edited by A. B. Rooth, pp. 33–63. Gleerups, Lund.

Hawkes, C. F., 1954, Archaeological Theory and Method: Some Suggestions from the Old World. *American Anthropologist* 56:155–168.

Hedrick, C. W., 1995, Thucydides and the Beginnings of Archaeology. In *Methods in the Mediterranean: Historical and Archaeological Views on Texts and Archaeology*, edited by D. B. Small, pp. 45–90. Mnemosyne, Supplement 135. Brill, Leiden, New York, and Cologne.

Heighway, C., and Bryant, R., in press, *The Anglo-Saxon Minster and Later Medieval Priory of St Oswald, Gloucester.* The Council for British Archaeology, London.

Hemer, O. (editor), 1994, *Kulturen i den globala byn.* Ægis, Lund.

Hesse B., 1995, Husbandry, Dietary Taboos and the Bones of the Ancient Near East: Zooarchaeology in the Post-Processual World. In *Methods in the Mediterranean: Historical and Archaeological Views on Texts and Archaeology*, edited by D. B. Small, pp. 197–232. Mnemosyne, Supplement 135. Brill, Leiden, New York, and Cologne.

Higham, C., 1989, *The Archaeology of Mainland Southeast Asia: From 10,000 B.C. to the Fall of Angkor.* Cambridge University Press, Cambridge.

Hildebrand, H., 1879–1903, *Sveriges medeltid I–V.* Norstedt, Stockholm.

Hildebrand, H., 1882, Historia och kulturhistoria. *Historisk Tidskrift* 2:1–28.

Hill, J. D., 1993, Can We Recognise a Different European Past? A Contrastive Archaeology of the Later Prehistoric Settlements in Southern England. *Journal of European Archaeology* 1:57–75.

Hingley, R., 1991, Past, Present and Future—The Study of the Roman Period in Britain. *Scottish Archaeological Review* 8:90–101.

Hinton, D. H. (editor), 1983, *25 Years of Medieval Archaeology*. The Department of Prehistory & Archaeology, University of Sheffield and The Society for Medieval Archaeology, Sheffield.

Hodder, I., 1982, *Symbols in Action: Ethnoarchaeological Studies of Material Culture*. Cambridge University Press, Cambridge.

Hodder, I., 1989, This Is Not an Article about Material Culture as Text. *Journal of Anthropological Archaeology* 8:250–269.

Hodder, I. (editor), 1991, *Archaeological Theory in Europe: The Last Three Decades*. Routledge, London and New York.

Hodder, I., 1994, Architecture and Meaning: The Example of Neolithic Houses and Tombs. In *Architecture and Social Space*, edited by M. Parker Pearson and C. Richards, pp. 73–86. Routledge, London and New York.

Hodder, I., and Shanks, M., 1995, Processual, Postprocessual and Interpretive Archaeologies. In *Interpreting Archaeology: Finding Meaning in the Past*, edited by I. Hodder, M. Shanks, A. Alexandri, V. Buchli, J. Carman, J. Last, and G. Lucas, pp. 3–29. Routledge, London and New York.

Hodges, R., 1982a, *Dark Age Economics: The Origins of Towns and Trade AD 600–1000*. Duckworth, London.

Hodges, R., 1982b, Method and Theory in Medieval Archaeology. *Archaeologia Medievale* 9:7–38.

Hodges, R., 1983, New Approaches to Medieval Archaeology, Part 2. In *25 Years of Medieval Archaeology*, edited by D. A. Hinton, pp. 24–32. The Department of Prehistory & Archaeology, University of Sheffield and The Society for Medieval Archaeology, Sheffield.

Hodges, R., 1989, *The Anglo-Saxon Achievement*. Duckworth, London.

Hoffmann, H., 1979, In the Wake of Beazley: Prolegomena to an Anthropological Study of Greek Vase-Painting. *Hephaistos* 1:61–70.

Hole, F. (editor), 1987, *The Archaeology of Western Iran: Settlement and Society from Prehistory to the Islamic Conquest*. Smithsonian Institution Press, Washington D.C.

Holl, A., 1990, West African Archaeology: Colonialism and Nationalism. In *A History of African Archaeology*, edited by P. Robertshaw, pp. 296–308. James Currey, London.

Honour, H., 1968, *Neo-Classicism*. Penguin, Harmondsworth.

Houby-Nielsen, S., 1997, Grave Gifts, Women and Conventional Values in Hellenistic Athens. In *Conventional Values of the Hellenistic Greeks*, edited by P. Bilde, T. Engberg-Pedersen, L. Hannestad, and J. Zahle, pp. 220–262. Aarhus University Press, Århus.

Hrbek, I., 1981, Written Sources from the Fifteenth Century Onwards. In *Unesco General History of Africa, I. Methodology and African Prehistory*, edited by J. Ki-Zerbo, pp. 114–141. Unesco, Paris and London.

Hsia, N., 1986, The Classification, Nomenclature, and Usage of Shang Dynasty Jades. In *Studies of Shang Archaeology: Selected Papers from the International Conference on Shang Civilization*, edited by K.-C. Chang, pp. 207–236. Yale University Press, New Haven.

Huffmann, T. N., 1986, Cognitive Studies of the Iron Age in Southern Africa. *World Archaeology* 18:84–95.

Humphreys, S. C., 1978, *Anthropology and the Greeks*. Routledge & Kegan Paul, London.

Hyslop, J., 1984, *The Inca Road System*. Academic Press, Orlando.

Hyslop, J., 1990, *Inka Settlement Planning*. University of Texas Press, Austin.

Ideström, L., 1983, Lunds residenstomter: En jämförelse mellan skriftligt och arkeologiskt källmaterial. Unpublished seminar paper, Institute of Archaeology, University of Lund, Lund.

Jankuhn, H., 1973, Umrisse einer Archäologie des Mittelalters. *Zeitschrift für Archäologie des Mittelalters* 1:9–19.

Jansen, W., 1984, Current Research Concerns in Medieval Archaeology in West Germany. In *Archaeological Approaches to Medieval Europe*, edited by K. Biddick, pp. 281–300. XVIII Medieval Institute Publications, Western Michigan University, Kalamazoo.

Jantzen, U., 1986, *Einhundert Jahre Athener Institut 1874–1974*. Philip von Zabern, Mainz.

Johannesson, K., 1990, *Retorik eller konsten att övertyga*. Norstedts, Stockholm.

Johannsen, H., 1992, Danmarks Kirker: Baggrund og historie. In *Från romanik till nygotik: Studier i kyrklig konst och arkitektur tillägnade Evald Gustafsson*, edited by M. Ullén, pp. 205–216. Sveriges Kyrkor/Riksantikvarieämbetet, Stockholm.

Johansen. A. B., 1974, *Forholdet mellom teori og data i arkeologi og andre erfaringsvitenskaper*. Arkeologiske Skrifter 1, Bergen.

Johnson, M. H., 1993, *Housing Culture: Traditional Architecture in an English Landscape*. University College London Press, London.

Jones, C., 1991, Cycles of Growth at Tikal. In *Classic Maya Political History: Hieroglyphic and Archaeological Evidence*, edited by T. P. Culbert, pp. 102–127. Cambridge University Press, Cambridge.

Junko, H., 1989, Japanese Archaeology and Society. *Archaeological Review from Cambridge* 8(1):36–45.

Kardulias, P. N., 1994, Towards an Anthropological Historical Archaeology in Greece. *Historical Archaeology* 28(3):39–55.

Kåring, G., 1992, *När medeltidens sol gått ned: Debatten om byggnadsvård i England, Frankrike och Tyskland 1815–1914*. Almqvist & Wiksell International, Stockholm.

Karsten, I. A., 1987, *Minnesmerket, en del av vår identitet: Gjenoppbygging, revaluering og regenerasjon av historiske byer i Polen og Tsjekkoslovakia*. Institutionen för konstvetenskap, Stockholms universitet, Stockholm.

Keightley, D. N. (editor), 1983, *The Origins of Chinese Civilization*. University of California Press, Berkeley and London.

Keller, C., 1978, *Arkeologi: virklighetsflukt eller samfunnsformning?* Universitetsforlaget, Oslo.

Keller, D. R., and Rupp, D. W. (editors), 1983, *Archaeological Survey in the Mediterranean Area*. British Archaeological Reports, International Series 155. Oxford.

Kemp, B. J., 1977, The Early Development of Towns in Egypt. *Antiquity* 51:185–200.

Kemp, B. J., 1984, In the Shadow of Texts: Archaeology in Egypt. *Archaeological Review from Cambridge* 3(2):19–28.

Kemp, B. J., 1989, *Ancient Egypt: Anatomy of a Civilization*. Routledge, London.

Kense, F. J., 1990, Archaeology in Anglophone West Africa. In *A History of African Archaeology*, edited by P. Robertshaw, pp. 135–154. James Currey, London.

Khaldûn, I., 1967, *An Introduction to History: The Muqaddimah*. Translated from Arabic by F. Rosenthal, abridged and edited by N. J. Dawood. Routledge & Kegan Paul, London and Henley.

Kidder, E., 1977, *Ancient Japan: The Making of the Past*. Elsevier-Phaidon, Oxford.

Kirch, P. V., 1986, *Island Societies: Archaeological Approaches to Evolution and Transformation*. Cambridge University Press, Cambridge.

Ki-Zerbo, J. (editor), 1981, *Methodology and African Prehistory: Unesco General History of Africa*. Unesco, Paris & London.

Klackenberg, H., 1986, Feodalism i Finnveden: Biskop Henrik och Berga. In *Medeltiden och arkeologin: Festskrift till Erik Cinthio*, pp. 339–361. Lund Studies in Medieval Archaeology 1, Lund.

Klackenberg, H., 1992, *Moneta Nostra: Monetariseringen i medeltidens Sverige*. Lund Studies in Medieval Archaeology 10. Almqvist & Wicksell International, Stockholm.

Kleinbauer, W. E., 1992, *Early Christian and Byzantine Architecture: An Annotated Bibliography and Historiography*. Reference Publications, Boston.

Klejn, L. S., 1977, A Panorama of Theoretical Archaeology. *Current Anthropology* 18: 1–42.

Klejn, L. S., 1993, To Separate a Centaur: On the Relationship of Archaeology and History in Soviet Tradition. *Antiquity* 67:339–348.

Klejnstrup-Jensen, P., 1978, Ændringer i arkæologi og anthropologi. *Hikuin* 4:11–26.

Klindt-Jensen, O., 1975, *A History of Scandinavian Archaeology*. Thames & Hudson, London.

Knudsen, B. M., 1982, Boder. *Fortid og Nutid* 19:440–478.

Kobyliński, Z., 1991, Theory in Polish Archaeology 1960–90: Searching for Paradigms. In *Archaeological Theory in Europe: The Last Three Decades*, edited by I. Hodder, pp. 223–247. Routledge, London and New York.

Kohl, P. L., 1989, The Use and Abuse of World Systems Theory: The Case of the "Pristine" West Asian State. In *Archaeological Thought in America*, edited by C. C. Lamberg-Karlovsky, pp. 218–240. Cambridge University Press, Cambridge.

Kosso, P., 1995, Epistemic Independence between Texual and Material Evidence. In *Methods in the Mediterranean: Historical and Archaeological Views on Texts and Archaeology*, edited by D. B. Small, pp. 177–196. Mnemosyne, Supplement 135. Brill, Leiden, New York, and Cologne.

Kotsakis, K., 1991, The Powerful Past: Theoretical Trends in Greek Archaeology. In *Archaeological Theory in Europe: The Last Three Decades*, edited by I. Hodder, pp. 65–90. Routledge, London and New York.

Krautheimer, R., 1942, Introduction to an Iconography of Mediaeval Architecture. *Journal of the Warburg and Courtauld Institutes* 5:1–33.

Krautheimer, R., 1965, *Early Christian and Byzantine Architecture*. Penguin, Harmondsworth.

Kristiansen, K., 1991, Prehistoric Migrations—The Case of the Single Grave and Corded Ware Cultures. *Journal of Danish Archaeology* 8:211–225.

Kuhn, T. S., 1970, *The Structure of Scientific Revolutions*, 2nd ed. University of Chicago Press, Chicago.

Kuper, A., 1988, *The Invention of Primitive Society: Transformations of an Illusion*. Routledge, London and New York.

Lagerroth, U.-B., Lund, H., Luthersson, P., and Mortenson, A. (editors), 1993, *I musernas tjänst: Studier i konstarternas interrelationer*. Brutus Östling förlag Symposion, Stockholm and Stehag.

Lal, M., 1984, *Settlement History and the Rise of Civilization in the Ganga—Yaruma Doab from 1500 BC–300 AD*. B R Publishing House, Delhi.

Lamberg-Karlovsky, C. C., 1989, Mesopotamia, Central Asia and the Indus Valley: So the Kings Were Killed. In *Archaeological Thought in America*, edited by C. C. Lamberg-Karlovsky, pp. 241–267. Cambridge University Press, Cambridge.

Larsen, M. T., 1984, Skriften som historiens motor? In *Skrift og Samfund*, pp. 5–23. Center for sammenlignende kulturforskning, temarapport 1, Københavns universitet, Copenhagen.

Larsen, M. T., 1987, Orientalism and the Ancient Near East. *Culture and History* 2: 96–115.

Larsen, M. T., 1988, Introduction: Literacy and Social Complexity. In *State and Society: The Emergence and Development of Social Hierarchy and Political Centralization*, edited by J. Gledhill, B. Bender, and M. T. Larsen, pp. 173–191. Unwin Hyman, London.

Larsen, M. T., 1994, *Sunkne paladser: Historien om Orientens opdagelse*. Gyldendal, Copenhagen.

Laurence, R., 1994, *Roman Pompeii: Space and Society*. Routledge, London and New York.

Leciejewicz, L., 1980, Medieval Archaeology and Its Problems. In *Unconventional Archaeology: New Approaches and Goals in Polish Archaeology*, edited by R. Schild, pp. 191–211. Polska Akademia Nauk, Wroclaw.

Leone, M. P., 1973, Archaeology as the Science of Technology: Mormon Town Plans and Fences. In *Research and Theory in Current Archaeology*, edited by Ch. L. Redman, pp. 125–150. Wiley, New York.

Leone, M. P., 1984, Interpreting Ideology in Historical Archaeology: Using the Rules of Perspective in the William Paca Garden in Annapolis, Maryland. In *Ideology, Power and Prehistory*, edited by D. Miller and Ch. Tilley, pp. 25–35. Cambridge University Press, Cambridge.

Leone, M. P., 1988, The Georgian Order as the Order of Merchant Capitalism in Annapolis, Maryland. In *The Recovery of Meaning: Historical Archaeology in the Eastern United States*, edited by M. P. Leone and P. B. Potter, pp. 235–261. Smithsonian Institution Press, Washington, D.C.

Leone, M. P., and Crosby, C. A., 1987, Middle-Range Theory in Historical Archaeology. In *Consumer Choice in Historical Archaeology*, edited by S. M. Spencer-Wood, pp. 397–410. Plenum Press, New York.

Leone, M. P., and Potter, P. B., 1988, Introduction: Issues in Historical Archaeology. In *The Recovery of Meaning: Historical Archaeology in the Eastern United States*, edited by M. P. Leone and P. B. Potter, pp. 1–22. Smithsonian Institution Press, Washington, D.C.

Le Vine, T. Y. (editor), 1992, *Inka Storage System*. University of Oklahoma Press, Norman and London.

Lewis, K. E., 1984, *The American Frontier: An Archaeological Study of Settlement Pattern and Process*. Academic Press, New York.

Liebgott, N.-K., 1989, *Dansk Middelalderarkæologi*. Gad, Copenhagen.

Liedman, S.-E., 1977, *Motsatsernas spel: Friedrich Engels' filosofi och 1800-talets vetenskaper 1-2*. Bo Cavefors, Lund.

Liedman, S.-E., 1978, Humanistiska forskningstraditioner i Sverige. Kritiska och historiska perspektiv. In *Humaniora på undantag? Humanistiska forskningstraditioner i Sverige*, edited by T. Forser, pp. 9–78. Pan/Norstedts, Stockholm.

Lindqvist, C., 1989, *Tecknens Rike: En berättelse om kineserna och deras skrivtecken*. Bonniers, Stockholm.

Little, B. J. (editor), 1992, *Text-Aided Archaeology*. CRC Press, Boca Raton.

Little, B. J., 1994, People with History: An Update on Historical Archaeology in the United States. *Journal of Archaeological Method and Theory* 1(1):5–40.

Lloyd, J. A., 1986, Why Should Historians Take Archaeology Seriously? In *Archaeology at the Interface: Studies in Archaeology's Relationships with History, Geography, Biology and Physical Science*, edited by J. L. Bintliff and C. F. Gaffney, pp. 40–51. British Archaeological Reports, International Series 300, Oxford.

Lloyd, S., 1980, *Foundations in the Dust: The Story of Mesopotamian Exploration*. Thames & Hudson, London.

Löfting, C., 1993, Imperiet i fransk selvopfattelse 1871–1962. In *Imperium Romanum: Realitet, idé, ideal III*, edited by S. O. Due and J. Isager, pp. 69–96. Sfinx, Århus.

Lovejoy, A. O., 1936, *The Great Chain of Being: A Study of the History of an Idea*. Harvard University Press, Cambridge, Massachusetts.

Lovejoy, P. E., 1986, Nigeria: The Ibadan School and Its Critics. In *African Historiographies: What History for Which Africa?* edited by B. Jewsiewicki and D. Newbury, pp. 197–205. Sage Publications, Beverly Hills, London, and New Delhi.

Lumbreras, L. G., 1974, *Arqueología como Ciencia Social*. Ediciones Hista, Lima.

Lund, A. A., 1993, *De etnografiske kilder til Nordens tidlige historie*. Aarhus universitetsforlag, Århus.

MacGaffey, W., 1986, Epistemological Ethnocentrism in African studies. In *African Historiographies: What History for Which Africa?* edited by B. Jewsiewicki and D. Newbury, pp. 42–48. Sage Publications, Beverly Hills, London, and New Delhi.

Mahler, D. L., Paludan-Müller, C., and Hansen, S. S., 1983, *Om Arkæologi: Forskning, formidling, forvaltning—for hvem?* Reitzel, Copenhagen.

Malcus, N., 1970, *Kina och Västerlandet: Attityder och påverkan från 1200-talet till 1949*. Akademiförlaget, Gothenburg.

Malik, S. C., 1975, *Understanding Indian Civilization: A Framework of Enquiry*. Indian Institute of Advanced Study, Simla.

Mallory, J. P., 1989, *In Search of the Indo-Europeans: Language, Archaeology and Myth*. Thames & Hudson, London.

Malmer, M. P., 1962, *Jungneolitische Studien*. Acta Archaeologica Lundensia, Series in 8° N° 2. CWK Gleerups, Lund.

Manatunga, A., 1994, Buddhist Epistemology and Archaeological reasoning: Some Correlations. Paper presented at the World Archaeological Congress 3, New Delhi, December 1994.

Mandal, D., 1993, *Ayodhya: Archaeology after Demolition: A Critique of the "New" and "Fresh" Discoveries*. Orient Longman, Hyderabad.

Marcus, J., 1976, *Emblem and State in the Classic Maya Lowlands*. Dumbarton Oaks, Washington D.C.

Marcus, J., 1992, *Mesoamerican Writing Systems: Propaganda, Myth, and History in Four Ancient Civilizations*. Princeton University Press, Princeton.

Matenga, E., 1993, *Archaeological Figurines from Zimbabwe*. Studies in African Archaeology 5. Uppsala.

Mathews, P., 1991, Classic Emblem Glyphs. In *Classic Maya Political History: Hieroglyphic and Archaeological Evidence*, edited by T. P. Culbert, pp. 19–29. Cambridge University Press, Cambridge.

Matos Mendieta, R. M., 1994, Peru: Some Comments. In *History of Latin American Archaeology*, edited by A. Oyuela-Caycedo, pp. 104–123. Avebury, Aldershot.

Matos Moctezuma, E., 1988, *The Great Temple of the Aztecs*. Thames and Hudson, London.

Matsumoto N., Nakazono, S., and Hodgkinson, L., 1994, The Formation Process and Current State of the Academically Isolated Archaeology of Japan. Paper presented at the World Archaeological Congress 3, New Delhi, December 1994.

McDonald, W. A., and Rapp, G. R., Jr. (editors), 1972, *The Minnesota Messenia Expedition: Reconstructing a Bronze Age Regional Environment*. The University of Minnesota Press, Minneapolis.

McGuire, R. H., 1991, Building Power in the Cultural Landscape of Broome County, New York 1880 to 1940. In *The Archaeology of Inequality*, edited by R. H. McGuire and R. Paynter, pp. 102–124. Blackwell, Oxford.

McGuire, R. H., 1992, *A Marxist Archaeology*. Academic Press, San Diego.

McGuire, R. H., and Paynter, R. (editors), 1991, *The Archaeology of Inequality*. Blackwell, Oxford.

McIntosh, S. K., and McIntosh, R. J., 1984, The Early City in West Africa: Towards an Understanding. *The African Archaeological Review* 2:73–98.

McIntosh, S. K., and McIntosh, R. J., 1993, Cities without Citadels: Understanding Urban Origins along the Middle Niger. In *The Archaeology of Africa: Food, Metals and Towns*, edited by T. Shaw, P. J. J. Sinclair, B. Andah, and A. Okpoko, pp. 622–641. Routledge, London and New York.

McNeal, R. W., 1972, The Greeks in History and Prehistory. *Antiquity* 46:19–28.

Meyer, W., 1985, *Hirsebrei und Hellebarde: Auf den Spuren des mittelalterlichen Lebens in der Schweiz*. Walter, Olten and Freiburg im Breisgau.

Miller, A. G., 1983, Image and Text in Pre-Hispanic Art: Apples and Oranges. In *Text and Image in Pre-Columbian Art: Essays on the Interrelationship of the Verbal and Visual Arts*, edited by J. C. Berlo, pp. 41–54. British Archaeological Reports, International Series 180, Oxford.

Miller, D., 1985, *Artefacts as Categories: A Study of Ceramic Variability in Central India*. Cambridge University Press, Cambridge.

Miller, J. M., 1991, Is It Possible to Write a History of Israel without Relying on the Hebrew Bible? In *The Fabric of History: Text, Artifacts and Israel's Past*, edited by D. V. Edelman, pp. 93–102. Journal for the Study of the Old Testament, Supplement Series 127. Sheffield Academic Press, Sheffield.

Mogren, M., 1990, Ökad utblick för ökad insikt—en slags inledning. *Meta* 90(4):2–16.

Moniot, H., 1986, Profile of a Historiography: Oral Tradition and Historical Research in Africa. In *African Historiographies: What History for Which Africa?* edited by B. Jewsiewicki and D. Newbury, pp. 50–58. Sage Publications, Beverly Hills, London, and New Delhi.

Montelius, O., 1900, Typologien eller utvecklingsläran tillämpad på det menskliga arbetet. *Svenska Fornminnesföreningens tidskrift* 10:237–268.

Moorey, R., 1991, *A Century of Biblical Archaeology*. The Lutterworth Press, Cambridge.

Moreland, J. F., 1991, Method and Theory in Medieval Archaeology in the 1990's. *Archaeologia Medievale* 18:7–42.

Moreland, J. F., 1992, Restoring the Dialectic: Settlement Patterns and Documents in Medieval Central Italy. In *Archaeology, Annals, and Ethnohistory*, edited by A. B. Knapp, pp. 112–129. Cambridge University Press, Cambridge.

Morris, C., 1988, Progress and Prospect in the Archaeology of the Inca. In *Peruvian Prehistory: An Overview of the Pre-Inca and Inca Society*, edited by R. W. Keatinge, pp. 233–256. Cambridge University Press, Cambridge.

Morris, C., 1992, Foreword. In *Inka Storage Systems*, edited by T. Y. LeVine, pp. xi–xiii. University of Oklahoma Press, Norman and London.

Morris, C., and Thompson, D. E., 1985, *Huánuco Pampa: An Inca City and Its Hinterland*. Thames and Hudson, London.

Morris, I., 1987, *Burial and Ancient Society: The Rise of the Greek City-State*. Cambridge University Press, Cambridge.

Morris, I., 1994, Archaeologies of Greece. In *Classical Greece: Ancient Histories and Modern Archaeologies*, edited by I. Morris, pp. 8–47. Cambridge University Press, Cambridge.

Morris, R., 1989, *Churches in the Landscape*. Dent & Sons, London.

Mosquera, G., 1994, Vid skiljevägen. In *Kulturen i den globala byn*, edited by O. Hemer, pp. 119–126. Ægis, Lund.

Munro-Hay, S., 1991, *Aksum: An African Civilization of Late Antiquity*. Edinburgh University Press, Edinburgh.

Murra, J. V., and Morris, C., 1975, Dynastic Oral Tradition, Administrative Records and Archaeology in the Andes. *World Archaeology* 7:269–279.

Murray, T., and Walker, M. J., 1988, Like WHAT? A Practical Question of Archaeological Inference and Archaeological Meaningfulness. *Journal of Anthropological Archaeology* 7:248–287.

Näsman, U., 1988, Analogislutning i nordisk jernalderarkæologi: Et bidrag til udviklingen af en nordisk historisk etnografi. In *Fra Stamme til Stat i Danmark 1: Jernalderens stammesamfund*, edited by P. Mortensen and B. M. Rasmussen, pp. 123–140. Jysk Arkæologisk Selskabs Skrifter 22, Århus.

Neale, C., 1986, The Idea of Progress in the Revision of African History, 1960–1970. In *African Historiographies: What History for Which Africa?* edited by B. Jewsiewicki and D. Newbury, pp. 112–122. Sage Publications, Beverly Hills, London, and New Delhi.

Nelson, S. M., 1993, *The Archaeology of Korea*. Cambridge University Press, Cambridge.

Netherly, P. J., 1988, From Event to Process: The Recovery of Late Andean Organizational Structure by Means of Spanish Colonial Written Records. In *Peruvian Prehistory: An Overview of the Pre-Inca and Inca Society*, edited by R. W. Keatinge, pp. 257–275. Cambridge University Press, Cambridge.

Netzer, E., 1991, The Buildings: Stratigraphy and Architecture. *Masada III: The Yigael Yadin Excavations 1963–1965, Final Reports*. Israel Exploration Society and The Hebrew University of Jerusalem, Jerusalem.

Niemeyer, H. G., 1968, *Einführung in die Archäologie*. Wissenschaftliche Buchgesellschaft, Darmstadt.

Nilsson, S. Å., 1968, *European Architecture in India, 1750–1850*. Faber and Faber, London.

Nissen, H. J., 1986, The Archaic Texts from Uruk. *World Archaeology* 17:317–334.

Noël Hume, I., 1969, *Historical Archaeology*. Knopf, New York.

Nordeide, S. W., 1989, Betente spor. *Meta* 1989 (1):31–46.

Nordenstam, T., 1993, *Från konst till vetenskap*. Carlsson, Stockholm.

Ober, J., 1995, Greek Horoi: Artifactual Texts and the Contingency of Meaning. In *Methods in the Mediterranean: Historical and Archaeological Views on Texts and Archaeology*, edited by D. B. Small, pp. 91–123. Mnemosyne, Supplement 135. Brill, Leiden, New York, and Cologne.

O'Connor, D., 1990, Egyptology and Archaeology: An African Perspective. In *A History of African Archaeology*, edited by P. Robertshaw, pp. 236–251. James Currey, London.

Oguagha, P. A., and Okpoko, A. I., 1984, *History and Ethnoarchaeology in Eastern Nigeria: A Study of Igbo–Igala Relations with Special Reference to the Anambra Valley*. British Archaeological Reports, International Series 195, Oxford.

Olsen, B., and Kobyliński, Z., 1991, Ethnicity in Anthropological and Archaeological Research: A Norwegian-Polish Perspective. *Archaeologia Polona* 29:5–27.

Ong, W., 1982, *Orality and Literacy: The Technologizing of the Word*. Methuen, London.

Orser, C. E., Jr., 1988a, *The Material Basis of the Postbellum Tenant Plantation. Historical Archaeology in the South Carolina Piedmont*. The University of Georgia Press, Athens, Georgia.

Orser, C. E., Jr., 1988b, Toward a Theory of Power for Historical Archaeology: Plantations and Space. In *Recovering of Meaning: Historical Archaeology in the Eastern United States*, edited by M. P. Leone and P. B. Potter, pp. 313–343. Smithsonian Institution Press, Washington, D.C.

Orser, C. E., Jr., 1996, *A Historical Archaeology of the Modern World*. Plenum Press, New York and London.

Orser, C. E., Jr., and Fagan, B. M., 1995, *Historical Archaeology*. Harper Collins, New York.

Österberg, E., 1978, *Historia och arkeologi: Några reflexioner*. Riksantikvarieämbetet och Statens historiska museer, Rapport 1978:2, Stockholm.

Oyuela-Caycedo, A. (editor), 1994, *History of Latin American Archaeology*. Avebury, Aldershot.

Paddayya, K., 1990, *The New Archaeology and Aftermath: A View from Outside the Anglo-American World*. Ravish Publishers, Pune.

Paddayya, K., 1995, Theoretical Perspectives in Indian Archaeology: An Historical Review. In *Theory in Archaeology: A World Perspective*, edited by P. J. Ucko, pp. 110–149. Routledge, London and New York.

Paine, R. T., and Soper, A., 1955, *The Art and Architecture of Japan*. Penguin, Harmondsworth.

Palaima T. G., and Shelmerdine, C. W., 1984, Mycenaean Archaeology and the Pylos Texts. *Archaeological Review from Cambridge* 3(2):76–89.

Palaima T. G., and Wright, J. C., 1985, Ins and Outs of the Archives Rooms at Pylos: Form and Function in a Mycenaean Palace. *American Journal of Archaeology* 89:251–262.

Panofsky, E., 1951, *Gothic Architecture and Scholasticism: An Inquiry into the Analogy of the Arts, Philosophy, and Religion in the Middle Ages*. Meridian, New York.

Pärssinen, M., 1992, *Tawantinsuyu: The Inca State and Its Political Organization*. Societas Historiae Finlandiae, Helsinki.

Patterson, T. C., 1989, Political Economy and a Discourse Called "Peruvian Archaeology." *Culture and History* 4:35–64.

Paynter, R., 1988, Steps to an Archaeology of Capitalism. In *The Recovery of Meaning: Historical Archaeology in the Eastern United States*, edited by M. P. Leone and P. B. Potter, pp. 407–433. Smithsonian Institution Press, Washington, D.C.

Pedersen, P., 1990, Classicism, Renaissance, and Post-Modernism. *Acta Hyperborea* 2:267–281.

Piggott, S., 1945, *Some Ancient Cities of India*. Geoffrey Cumberlege, Oxford.

Pikirayi, I., 1993, *The Archaeological Identity of the Mutapa State: Towards an Historical Archaeology of Northern Zimbabwe*. Studies in African Archaeology 6, Uppsala.

Pirenne, H., 1925, *Medieval Cities: Their Origins and the Revival of Trade*. Princeton University Press, Princeton, New Jersey.

Platzack, S., 1979, *Språkvetenskapens historia*. Pankosmion House, Athens.

Politis, G., 1995, The Socio-Politics of the Development of Archaeology in Hispanic South America. In *Theory in Archaeology: A World Perspective*, edited by P. J. Ucko, pp. 236–250. Routledge, London and New York.

Postgate, J. N., 1984, Cuneiform Catalysis: The First Information Revolution. *Archaeological Review from Cambridge* 3(2):4–18.

Postgate, J. N., 1990, Archaeology and the Texts—Bridging the Gap. *Zeitschrift für Assyriologie* 80:228–240.

Postgate, J. N., 1992, *Early Mesopotamia: Society and Economy at the Dawn of History*. Routledge, London and New York.

Potter, T. W., 1979, *The Changing Landscape of South Etruria*. Paul Elek, London.

Potter, T. W., 1987, *Roman Italy*. British Museum Publications, London.

Potts, D. T., 1990, *The Arabian Gulf in Antiquity 1–2*. Clarendon Press, Oxford.

Rahtz, P., 1981, *The New Medieval Archaeology*. Inaugural lecture delivered at the University of York, November 12, 1980, York.

Rahtz, P., 1983, New Approaches to Medieval Archaeology, Part 1. In *25 Years of Medieval Archaeology*, edited by D. A. Hinton, pp. 12–23. The Department of Prehistory & Archaeology, University of Sheffield and The Society for Medieval Archaeology, Sheffield.

Randsborg, K., 1980, *The Viking Age in Denmark: The Formation of a State*. Duckworth, London.

Randsborg, K., 1993, Kivik: Archaeology and Iconography. *Acta Archaeologica* 64:1–147.

Randsborg, K., 1994, Ole Worm: An Essay on the Modernization of Antiquity. *Acta Archaeologica* 65:135–169.

Rao, N., 1994, Symbol and History in Ram Janmabhoomi/Babri Masjid. In *Social Construction of the Past: Representation as Power*, edited by G. C. Bond and A. Gilliam, pp. 154–164. Routledge, London and New York.

Rathje, A., and Lund, J., 1991, Danes Overseas—A Short History of Danish Classical Archaeological Fieldwork. *Acta Hyperborea* 3:11–56.

Rathje, W., and Murphy, C., 1992, *Rubbish! The Archaeology of Garbage*. Harper Perennial, New York.

Rautman M. L., 1990, Archaeology and Byzantine Studies. *Byzantinische Forschungen* 15:137–165.

Ravn, M., 1993, Analogy in Danish Prehistoric Studies. *Norwegian Archaeological Review* 26:59–75.

Redin, L., 1977, *Lagmanshejdan. Ett gravfält som spegling av sociala strukurer i Skanör*. Acta Archaeologica Lundensia, Series in 4° N° 10. Gleerups, Lund.

Redin, L., 1982, Stadsarkeologi—perfektum i viss mån presens och något futurum. *Bebyggelsehistorisk tidskrift* 3:21–30.

Redman, C. L., 1986, *Qsar es-Seghir: An Archaeological View of Medieval Life*. Academic Press, Orlando.

Reid, D. M., 1985, Indigenous Egyptology: The Decolonization of a Profession. *Journal of the American Oriental Society* 105:233–246.

Renfrew, C., 1972, *The Emergence of Civilization: The Cyclades and the Aegean in the Third Millenium BC*. Methuen, London.

Renfrew, C., 1980, The Great Tradition versus the Great Divide: Archaeology as Anthropology. *American Journal of Archaeology* 84:287–298.

Renfrew, C., 1987, *Archaeology and Language: The Puzzle of Indo-European Origins*. Jonathan Cape, London.

Renfrew, C., and Cherry, J. F. (editors), 1986, *Peer Polity Interaction and Sociopolitical Change*. Cambridge University Press, Cambridge.

Rhodes, C., 1994, *Primitivism and Modern Art*. Thames & Hudson, London

Robertshaw, P. (editor), 1990a, *A History of African Archaeology*. London: James Currey, London.

Robertshaw, P., 1990b, Development of Archaeology in East Africa. In *History of African Archaeology*, edited by P. Robertshaw, pp. 78–94. James Currey, London.

Rodwell, W., 1981, *Church Archaeology*. Batsford, London.

Ruan, Y., 1993, The Conservation of Chinese Historic Cities. *Antiquity* 67:850–856.

Rubin, W. (editor), 1984, *Primitivism in the 20th Century Art: Affinity of the Tribal and the Modern*. Museum of Modern Art, New York.

Rumpf, A., 1953, *Archäologie. I Einleitung, Historischer Überblick*. de Gruyter, Berlin.

Said, E. W., 1978, *Orientalism: Western Conceptions of the Orient*. Routledge & Kegan Paul, London.

Samson, R., 1987, Social Structures from Reihengräber: Mirror or Mirage? *Scottish Archaeological Review* 4(2):116–126.

Sanders, W. T., and Webster, D. W., 1988, The Mesoamerican Urban Tradition. *American Anthropologist* 90:521–546.

Sanders, W. T., Parsons, J. R., and Santley, R. S., 1978, *The Basin of Mexico: Ecological Processes in the Evolution of the Civilization*. Academic Press, New York.

Sawyer, P. H., 1983, English Archaeology before the Conquest: A Historian's View. In *25 Years of Medieval Archaeology*, edited by D. A. Hinton, pp. 44–47. The Department of Prehistory & Archaeology, University of Sheffield and The Society for Medieval Archaeology, Sheffield.

Schaedel, R. P., 1992, The Archaeology of the Spanish Colonial Experience in South America. *Antiquity* 66:217–242.

Schávelzon, D., 1989, The History of Mesoamerican Archaeology at the Crossroad: Changing Views of the Past, In *Tracing Archaeology's Past: The Historiography of Archaeology*, edited by A. L. Christenson, pp. 107–112. Southern Illinois University Press, Carbondale and Edwardsville.

Schiering, W., 1969, Zur Geschichte der Archäologie. In *Allgemeine Grundlagen der Archäologie, Begriff und Methode, Geschichte, Problem der Form, Schriftzeugnisse*, edited by U. Hausmann, pp. 11–161. Handbuch der Archäologie, H. C. Beck, Munich.

Schliemann, H., 1878, *Mykenae: Bericht über meine Forschungen und Entdeckungen in Mykenae und Tiryns*. Brockhaus, Leipzig.

Schmandt-Besserat, D., 1992, *Before Writing I: From Counting to Cuneiform*. University of Texas Press, Austin.

Schmidt, P. R., 1978, *Historical Archaeology: A Structural Approach in an African Culture*. Greenwood Press, Westport.

Schmidt, P. R., 1983, An Alternative to a Strictly Materialist Perspective: A Review of Historical Archaeology, Ethnoarchaeology, and Symbolic Approaches in African Archaeology. *American Antiquity* 48:62–79.

Schmidt, P. R., 1990, Oral Traditions, Archaeology and History: A Short Reflective History. In *A History of African Archaeology*, edited by P. Robertshaw, pp. 252–270. James Currey, London.

Schnapp, A., 1993, *La conquête du passé: Aux origines de l'archéologie*. Éditions Carré, Paris.

Schreiber, K. J., 1992, *Wari Imperialism in Middle Horizon Peru*. Anthropological Papers, Museum of Anthropology, University of Michigan, Ann Arbor.

Schrire, C., 1991, The Historical Archaeology of the Impact of Colonialism in Seventeenth-Century South Africa, In *Historical Archaeology in Global Perspective*, edited by L. Falk, pp. 69–96. Smithsonian Institution Press, Washington, D.C.

Schuyler, R. L., 1968, The Use of Historic Analogy in Archaeology. *American Antiquity* 33:390–392.

Schuyler, R. L., 1970, Historical Archaeology and Historic Sites Archaeology as Anthropology: Basic Definitions and Relationships. *Historical Archaeology* 4:83–89.

Schuyler, R. L. (editor), 1978 *Historical Archaeology: A Guide to Substantive and Theoretical Contributions*. Baywood, Farmingdale.

Schuyler, R. L. (editor), 1980, *Archaeological Perspectives on Ethnicity in America: Afro-American and Asian American Culture History*. Baywood Monographs in Archaeology 1. Baywood, Farmingdale.

Schuyler, R. L., 1988, Archaeological Remains, Documents, and Anthropology: A Call for a New Culture History. *Historical Archaeology* 22(1):36–42.

Scott, E., 1993, Writing the Roman Empire. In *Theoretical Roman Archaeology: First Conference Proceedings*, edited by E. Scott, pp. 5–22. Avebury, Aldershot,

Seidenspinner, W., 1989, Mittelalterarchäologie und Volkskunde: Ein Beitrag zur Öffnung und zur Theoriebildung archäologischer Mittelalterforschung. *Zeitschrift für Archäologie des Mittelalters* 14–15:9–48.

Seifert, D. J. (editor), 1991, Gender in Historical Archaeology. *Historical Archaeology* 25:4.

Serlio, S., 1609, *Von der Architectur*. Ludwig König, Basle.

Shackel, P. A., 1993, *Personal Discipline and Material Culture: An Archaeology of Annapolis, Maryland*. University of Tennessee Press, Knoxville.

Shanks, M., and Tilley, C., 1987, *Re-Constructing Archaeology: Theory and Practice*. Cambridge University Press, Cambridge.

Sharer, R. J., 1991, Diversity and Continuity in Maya Civilization: Quirigara as a Case

Study. In *Classic Maya Political History: Hieroglyphic and Archaeological Evidence*, edited by T. P. Culbert, pp. 180–198. Cambridge University Press, Cambridge.

Sharma, R. S., 1987, *Urban Decay in India (c. 300–c. 1000)*. Munshiram Manoharlal, New Delhi.

Sharma, R. S., 1990, Keynote Address. In *Historical Archaeology in India: A Dialogue between Archaeologists and Historians*, edited by A. Ray and S. Mukherjee, pp. 1–11. Books & Books, New Delhi.

Shaw, T., 1989, African Archaeology: Looking Back and Looking Forward. *The African Archaeological Review* 7:3–131.

Shaw, T., Sinclair, P. J. J., Andah, B., and Okpoko, A. (editors), 1993, *The Archaeology of Africa: Food, Metals and Towns*. Routledge, London and New York.

Shay, T., 1989, Israeli Archaeology—Ideology and Practice. *Antiquity* 63:768–772.

Silberman, N. A., 1982, *Digging for God and Country: Exploration, Archaeology and the Secret Struggle for the Holy Land 1799–1917*. Knopf, New York.

Silberman, N. A., 1989, *Between Past and Present: Archaeology, Ideology and Nationalism in the Modern Middle East*. Henry Holt, New York.

Sinclair, P. J. J., 1987, *Space, Time and Social Formation: A Territorial Approach to the Archaeology and Anthropology of Zimbabwe and Mozambique c. 0–1700 AD*. Aun 9, Uppsala.

Singleton, T. (editor), 1985, *The Archaeology of Slavery and Plantation Life*. Academic Press, New York.

Skydsgaard, J. E., 1970, *Pompeii: En romersk provinsby*. Gad, Copenhagen.

Small, D. B. (editor), 1995, *Methods in the Mediterranean: Historical and Archaeological Views on Texts and Archaeology*. Mnemnosyne, Supplement 135. Brill, Leiden, New York and Cologne.

Smith, R. B., and Watson, W. (editors), 1979, *Early South East Asia: Essays in Archaeology, History and Historical Geography*. Oxford University Press, New York and Oxford.

Snodgrass, A. M., 1983, Archaeology. In *Sources for Ancient History*, edited by M. Crawford, pp. 137–184. Cambridge University Press, Cambridge.

Snodgrass, A. M., 1985a, Greek Archaeology and Greek History. *Classical Antiquity* 4(2):193–207.

Snodgrass, A. M., 1985b, The New Archaeology and the Classical Archaeologist. *American Journal of Archaeology* 89:31–37.

Snodgrass, A. M., 1987, *An Archaeology of Greece: The Present State and Future Scope of a Discipline*. University of California Press, Los Angeles and London.

Sonesson, G., 1992, *Bildbetydelser: Inledning till bildsemiotiken som vetenskap*. Studentlitteratur, Lund.

South, S., 1977, *Method and Theory in Historical Archaeology*. Academic Press, New York.

Spencer-Wood, S., (editor), 1987, *Consumer Choice in Historical Archaeology*. Plenum Press, New York.

Spencer-Wood, S., 1994, Feminist Historical Archaeology in the USA. Paper presented at the World Archaeological Congress 3, New Delhi, December 1994.

Stahl, A. B., 1993, Concepts of Time and Approaches to Analogical reasoning in Historical Perspective. *American Antiquity* 58:235–260.

Steuer, H., 1982, *Frühgeschichtliche Sozialstrukturen in Mitteleuropa: Eine Analyse der Auswertungsmethoden des archäologischen Quellenmaterials*. Abhandlungen der Akademie der Wissenschaften in Göttingen. Philologisch-historische Klasse 3:128, Göttingen.

Steuer, H., 1989, Archaeology and History: Proposals on the Social Structure of the Merovingian Kingdom. In *The Birth of Europe: Archaeology and Social Development in the First Millennium* A.D., edited by K. Randsborg., pp. 100–122. Analecta Romana Instituti Danici, Supplementum XVI. L'Erma de Bretschneider, Rome.

Stjernquist, B., 1983, Sven Nilsson som banbrytare i svensk arkeologi. In *Sven Nilsson: En lärd i 1800-talets Lund*, edited by G. Regnéll, pp. 157–211. Kungliga Fysiografiska Sällskapet i Lund, Lund.

Stoianovich, T., 1976, *French Historical Method: The Annales Paradigm*. Cornell University Press, Ithaca and London.

Stone, E. C., 1987, *Nippur Neighborhood*. Studies in Ancient Oriental Civilization No. 44. The Oriental Institute of the University of Chicago, Chicago.

Stone, G. W., 1989, Artifacts Are Not Enough. In *Documentary Archaeology in the New World*, edited by M. C. Beaudry, pp. 68–78. Cambridge: Cambridge University Press, Cambridge.

Stoneman, R., 1987, *Land of Lost Gods: The Search for Classical Greece*. University of Oklahoma Press, Norman and London.

Strömbäck, D., 1971, Bröderna Grimm och folkminnesforskningens vetenskapliga grundläggning. In *Folkdikt och folktro*, edited by A. B. Rooth, pp. 19–32. Gleerups, Lund.

Stuart, D., 1988, The Rio Azul Cacao Pot: Epigraphic Observations on the Function of a Maya Ceramic Vessel. *Antiquity* 62:153–157.

Stuart, G. E., 1992, Quest for Decipherment: A Historical and Biographical Survey of Maya Hieroglyphic Investigation. In *New Theories on the Ancient Maya*, edited by E. C. Damien and R. J. Sharer, pp. 1–63. The University Museum, University of Pennsylvania, Philadelphia.

Stuart, J., and Revett, N., 1829–1833, *Die Alterthümer zu Athen*. Leske, Leipzig and Darmstadt.

Svenbro, J., 1989, Phrasikleia—An Archaic Greek Theory of Writing. In *Literacy and Society*, edited by K. Schousboe and M. T. Larsen, pp. 229–246. Akademisk Forlag, Copenhagen.

Svensson, S., 1966, *Introduktion till folklivsforskningen*. Natur och Kultur, Stockholm.

Tabaczyński, S., 1987, Medieval Archaeology: Problems. Sources. Methods. Task of Research. English summary of *Archeologia Srednowieczna: Problemy. Zródla. Metody. Cele Badawcze*, pp. 272–274. Ossolineum, Wroclaw.

Tabaczyński, S., 1993, The Relationship between History and Archaeology: Elements of the Present Debate, *Medieval Archaeology* 37:1–14.

Tanaka, M., 1987, The Early Historic Periods. In *Recent Archaeological Discoveries in Japan*, edited by K. Tsuboi, pp. 72–91. Unesco, Paris and The Centre for East Asian Cultural Studies, Tokyo.

Thomas, R., 1992, *Literacy and Orality in Ancient Greece*. Cambridge University Press, Cambridge.

Thomsen, C. J., 1836, Kortfattet Udsigt over Mindesmærker og Oldsager fra Nordens fortid. In *Ledetraad til Nordisk Oldkyndighed*, pp. 27–87. Det kongelige Nordiske Oldskrift-Selskab, Copenhagen.

Tilley, C. (editor), 1990, *Reading Material Culture: Structuralism, Hermeneutics and Poststructuralism*. Blackwell, Oxford.

Tilley, C., 1991, *Material Culture and Text: The Art of Ambiguity*. Routledge, London and New York.

Toulmin, S., and Goodfield, J., 1965, *The Discovery of Time*. Hutchinson, London.

Townsend, R. F., 1992, *The Aztecs*. Thames & Hudson, London.

Trigger, B. G., 1984, Alternative Archaeologies: Nationalist, Colonialist, Imperialist. *Man* 19:355–370.

Trigger, B. G., 1989, *A History of Archaeological Thought*. Cambridge University Press, Cambridge.

Trigger, B. G., 1990, The History of African Archaeology in World Perspective. In *A History of African Archaeology*, edited by P. Robertshaw, pp 309–319. James Currey, London.

Trigger, B. G., 1993, *Early Civilizations: Ancient Egypt in Context*. The American University of Cairo Press, Cairo.

Tsude, H., 1990, Chiefly Lineages in Kofun-Period Japan: Political Relations between Centre and Region. *Antiquity* 64:923–931.

Tsude, H., 1995, Archaeological Theory in Japan. In *Theory in Archaeology: A World Perspective*, edited by P. J. Ucko, pp. 298–311. Routledge, London and New York.

Unnerbäck, A., 1992, Sveriges Kyrkor—konsthistoriskt inventarium. In *Från romanik till nygotik: Studier i kyrklig konst och arkitektur tillägnade Evald Gustafsson*, edited by M. Ullén, pp. 195–204. Sveriges Kyrkor/Riksantikvarieämbetet, Stockholm.

Vansina, J., 1965, *Oral Tradition: A Study in Historical Methodology*. Routledge & Kegan Paul, London.

Vansina, J., 1984, *Art History in Africa*. Longman, London and New York.

Varga, A. K., 1989, Criteria for Describing Word-and-Image Relations. *Poetics Today* 10(1):31–53.

Vázquez León, L., 1994, Mexico: The Institutionalization of Archaeology, 1885–1942. In *History of Latin American Archaeology*, edited by A. Oyuela-Caycedo, pp. 69–89. Avebury, Aldershot.

Vickers, M., 1990, The Impoverishment of the Past: The Case of Classical Greece. *Antiquity* 64:455–463.

von Falkenhausen, L., 1993, On the Historiographical Orientation of Chinese Archaeology. *Antiquity* 67:839–849.

Wainwright, F. T., 1962, *Archaeology and Place-Names and History. An Essay on Problems of Co-ordination*. Routledge & Kegan Paul, London.

Wallace-Hadrill, A., 1994, *Houses and Society in Pompeii and Herculaneum*. Princeton University Press, Princeton, New Jersey.

Wapnish, P., 1995, Towards Establishing a Conceptual Basis for Animal Categories in Archaeology. In *Methods in the Mediterranean: Historical and Archaeological Views on Texts and Archaeology*, edited by D. B. Small, pp. 233–273.

Washburn, D. K., 1983, Symmetry Analysis of Ceramic Design: Two Tests of the Method on Neolithic Material from Greece and the Aegean. In *Structure and Cognition in Art*, edited by D. K. Washburn, pp. 138–164. Cambridge University Press, Cambridge.

Weiss, R., 1969, *The Renaissance Discovery of Classical Antiquity*. Blackwell, Oxford.

Welinder, S., 1991, The Word Förhistorisk, 'Prehistoric' in Swedish. *Antiquity* 65:295–296.

Welinder, S., 1992a, *Människor och landskap*. Aun 15, Uppsala.

Welinder, S., 1992b, *Människor och artefaktmönster*. Occasional Papers in Archaeology 5, Uppsala.

Welinder, S., 1994, The Ethnoarchaeology of a Swedish Village. *Current Swedish Archaeology* 2:195–209.

Wenke, R. J., 1989, Egypt: Origins of Complex Societies. *Annual Review of Anthropology* 18:129–155.

Wenskus, R., 1979, Randbemerkungen zum Verhältnis von Historie und Archäologie, insbesondere mittelalterliche Geschichte und Mittelalterarchäologie. In *Geschichtswissenschaft und Archäologie: Untersuchungen zur Siedlungs-, Wirtschafts- und Kirchengeschichte*, edited by H. Jankuhn and R. Wenskus, pp. 637–657. Vorträge und Forschungen 22, Sigmaringen.

Wheatley, P., 1971, *The Pivot of the Four Quarters: A Preliminary Enquiry into the Origins and Character of the Ancient Chinese City*. Edinburgh University Press, Edinburgh.

White, H., 1987, *The Content of the Form: Narrative Discourse and Historical Representation*. The Johns Hopkins University Press, Baltimore.

Wienberg, J., 1988, Metaforisk arkæologi og tingenes sprog. *Meta* 1988 (1–2):30–57.

Wienberg, J., 1993, *Den gotiske labyrint: Middelalderen og kirkerne i Danmark*. Lund Studies in Medieval Archaeology 11. Almqvist & Wiksell International, Stockholm.

Wilk, R., and Schiffer, M. B., 1981, The Modern Material Culture Field School: Teaching Archaeology on the University Campus. In *Modern Material Culture: The Archaeology of Us*, edited by R. A. Gould and M. B. Schiffer, pp. 15–30. Academic Press, New York.

Willey, G. R., and Sabloff, J. A., 1974, *A History of American Archaeology*. Thames & Hudson, London.

Wolf, E. R., 1982, *Europe and the People without History*. University of California Press, Berkeley.

Woolf, G., 1992, Imperialism, Empire and the Integration of the Roman Economy. *World Archaeology* 23(3):283–293.

Wurst, L., 1991, "Employees Must Be of Moral and Temperate Habits": Rural and Urban Elite Ideologies. In *The Archaeology of Inequality*, edited by R. H. McGuire and R. Paynter, pp. 125–149. Blackwell, Oxford.

Wylie, A., 1985, The Reaction against Analogy. In *Advances in Archaeological Method and Theory*, vol. 8, edited by M. B. Schiffer, pp. 63–111. Academic Press, Orlando.

Yamamoto, T., 1986, Reflections on the Development of Historical Archaeology in Japan. In *Windows on the Japanese Past: Studies in Archaeology and Prehistory*, edited by R. J. Pearson, pp. 397–403. Center for Japanese Studies, University of Michigan, Ann Arbor.

Yentsch, A. E., 1989, Farming, Fishing, Whaling, Trading: Land and Sea Resource on Eighteenth-Century Cape Cod. In *Documentary Archaeology in the New World*, edited by M. C. Beaudry, pp. 138–160. Cambridge University Press, Cambridge.

Yentsch, A. E., 1994, *A Chesapeake Family and Their Slaves: A Study in Historical Archaeology*. Cambridge University Press, Cambridge.

Young, R., 1990, *White Mythologies: Writing History and the West*. Routledge, London and New York.

Zanker, P., 1988a, *The Power of Images in the Age of Augustus*. University of Michigan Press, Ann Arbor.

Zanker, P., 1988b, *Pompeji: Stadtbilder als Spiegel von Gesellschaft und Herrschaftsform*. Philipp von Zabern, Mainz.

Index

Adams, Robert McC., 47
African anthropology, 74–75
African archaeology, 73, 75–82 120, 162
  colonialism, 74, 79–80
  excavations, 75, 80–82
  historical, 79–82
  historiography, 79–80
  professionalization of, 75
Aix-la-Chapelle, 10
Akhenaton, 43
Aksum Empire, 78
Alberti, Leon Battista, 108, 161
  De re ædificatoria, 108
Albright, William, 50
Alexander the Great, 38, 58
Andah, Bassey W., 77
Annales school, 32, 34, 122
Anyang: *see* Yinxu
Aphrodite, 175
Archaeology
  and analogy, 3, 131–132, 134, 156, 171
  Arabian, 37
  and association, 168–169, 171, 176
    chronological, 168
    spatial, 168
  Byzantine, 9
  Chinese: *see* Chinese archaeology
  classical: *see* Classical archaeology
  and classification, 157, 176
  and context, 24, 34, 67, 113, 127, 130, 155
  and contrast, 171, 176
  and correspondence, 176
  Egyptian: *see* Egyptology
  ethno-, 132, 148–149, 158
    historical, 162
  and excavations, 15, 17, 30, 132, 168
  and gender, 99
  historical: *see* Historical archaeology
  Indian: *see* Indian archaeology
  industrial, 9, 111
  Islamic, 37
  Israeli, 50

Archaeology (*cont.*)
  Japanese: *see* Japanese archaeology
  landscape, 21, 47
  marine, 9
  Marxist, 94, 122
  medieval: *see* Medieval archaeology
  Mesopotamian: *see* Mesopotamian archaeology
  Mexican: *see* Mexican archaeology
  New, 53, 77, 132
  and oral tradition, 81
  Peruvian: *see* Peruvian archaeology
  postmedieval, 122, 183
  postprocessual, 2, 6, 60, 132, 155
  prehistoric, 1–2, 32, 117, 122, 131, 136–138, 140, 142, 157, 176–177, 181
  processual, 132
  reconstruction, 111
  Russian, 122
  settlement, 32, 121
  Syro-Palestine, 53
  and tradition, 2
  and typology, 2, 26–27, 136, 157–158
Archaeology in the U.S., 97
  excavations, 97–98
  professionalization of, 97
Arikamedu, 58
Arnold, C. J., 132
Arrow War, 63
Arthasastra, 60–61
Artifacts and text, 3–4, 7–11, 13–14, 23, 25, 35, 47, 49, 60, 62, 67, 79, 90–91, 95, 102, 106–108, 117–118, 130, 135, 142, 144–145, 148–149, 151, 153, 155–156, 158, 161, 163–164, 166, 171, 175, 177, 180–181, 183
  and categories, 158
  and context, 155
  and correspondence, 157
  definition of, 146

Artifacts and text (*cont.*)
  and discursive contexts, 149, 155
    integrated, 151–152
    object-created, 150–151
    text-created, 152–153
  and documents, 148
  and objects, 147
Aryans, 60
Asiatic Society, 55
Asóka, 56–57, 61
Assyriology, 37, 180
Aswan Project, 41
Athens
  Agora, 18
  Kerameikos cemetery, 174
Attica, 24–25
Aubrey, John, 26
Augustus, 19
Austin, David, 33
Ayodhya, 59
Aztec Empire, 83, 86, 108
  Aztecs, 84, 87

Babylon, 44, 47, 165
  Ishtar Gate, 46
Bagley, R. W., 67
Baines, John, 42–43, 150
Barnard, Noël, 67
Barnes, Gina L., 72
Baudin, Louis, 95
Beaudry, Mary C., 102, 130, 134, 180
Beazly, John, 19
Bennet, John, 24
Berlin, 14
  Altes Museum, 14
Berlo, Janet Catherine, 91
Bernal, Ignacio, 90
Bianchi Bandinelli, Ranuccio, 17, 19,
    117
Bible, 45, 49–50, 53, 82
Biblical archaeology, 37, 38, 49–53, 117,
    121
  in Europe, 50, 52
  and landscape, 53
  professionalization, 50
  secularization of, 53
  in the United States, 50, 52
Biondo, Flavio, 120
Birchbark letters, 116, 169
Bloch, Marc, 122
Boer settlements, 78–79
de Bouard, Michel, 34, 105

Brahmi, 56
Braque, Georges, 74
Braudel, Fernand, 34
Bruneau, Philippe, 23
Brunius, Carl Georg, 27–28
Bruno, Giordano, 82
Buddha (Prince Siddharta Gautama), 58
Budé, Guillaume, 117
Burckhardt, Jacob, 128
Bureus, Johannes, 114
  Monumenta Sveo-Gothica, 114

Cairo, 40
Cape Colony, 79, 174
Cape Town, 79, 174
Catherwood, Frederick, 84
Cave paintings, 149
de Caylus, Count, 13
Chakrabati, Dilip, 56, 60
Champollion, Jean François, 39
Chang, K.-C., 67
Charles the Great, 10
Charlton, Thomas C., 132
Charnay, Désiré, 85
Chaze, Diane Z., 90
Childe, Gordon, 94, 122
Chinese archaeology, 62–68, 117, 123,
    139
  context, 67
  landscape, 65
  professionalization of, 63
Christopersen, Axel, 35, 180
Cinthio, Erik, 31
Clanchy, Michael, 147, 152
Clarke, David, 132
  The Loss of Innocence, 132
Classical archaeology, 1, 9–10–11, 15, 17,
    20–25, 28–29, 37, 107–108, 111,
    116–117, 121, 135, 138, 143,
    179–180
  in Australia, 10
  and classicism, 20
  in Europe, 10
  and future, 20–25
  in Greece, 10
  in Italy, 10, 19
  and landscape archaeology, 21
  and professionalization, 10, 15
  in the United States, 10
Colonialism, 74, 95, 97, 99, 111, 130,
    158
Confucius, 65

Constantinople, 38
  Hagia Sophia, 151
Conze, Alexander, 15, 17,105
Coomaraswany, Ananda K., 57, 60
Copán, 170
Corinth, 18
Cortéz, Hernán, 83
Cosmology
  biblical, 136–137
  evolutionary, 137, 139
  secular, 139, 141
Crosby, Constance A., 102, 171
Cultural Revolution, 66, 110
Cuneiform texts, 114, 116, 146, 149, 151,
  169
Cunningham, Alexander, 56, 121, 164
Curman, Sigurd, 29
Curtius, Ernst, 116
Cuzco, 95

Dark, K. R., 2
Darwin, Charles, 44
  On the Origin of Species, 44
Deagan, Kathleen, 102
Deetz, James, 130, 132–133
Dehio, George, 29
Delitzsch, Friedrich, 45
  Babel und Bibel, 45, 50
Delphi, 17
Dendrochronology, 160
Dethlefsen, Erwin, 132–133
Dever, William G., 53
Diderot, Denis, 162
  Encyclopédie, 162
Doppler effect, 132
Doré, Gustave, 51
Dorians, 17
Dörpfeld, Wilhelm, 15, 137
Driscoll, Stephen T., 35, 148
Driver, S. R., 50
Dymond, D. P., 35, 147
Dyson, Stephen L., 23

Egyptology, 1, 35, 38–43, 108, 117, 180
  and Britain, 39
  and France, 38–40
  landscape archaeology, 41
  professionalization of, 40–41
Elgin, James Bruce, Lord, 14
Eliot Smith, G., 38
Ellis, Maria de Jong, 49
Erdosy, George, 61

Eurocentrism, 142–143
Evans, Arthur, 18
Evolutionism, 138, 140
  and synchronism, 141
Expédition scientifique de Morée, 15

von Falkenhausen, Lothar, 65, 68
Fascism, 19
Fergusson, James, 56
Finley, Moses, 23, 126, 145
Fiorelli, Giuseppi, 15, 137
Flinders Petrie, William, 40, 50, 93
Florence, 11
Fort Necessity, 98
Fourth of May Movement, 63
Franken, H. J., 53
French Revolution, 110
Freud, Sigmund, 75

Gambio, Mañuel, 84
Gandhara, 55
Ganges, 58–60
Genesis, 1
Gerhard, Eduard, 15, 117
Giddens, Anthony, 130
Gilchrist, Roberta, 34
Glassie, Henry, 130
Gloucester, 112
  St Oswalds Church, 112
Graffiti, 149, 169
Greece, ancient, 11–25

Hall, Martin, 79, 143, 172, 174
Hammurabi, 171
Harvey, David, 141
Hawkes, Christopher, 124
Hebrew, rabbinical, 147
Herculaneum, 11
von Herder, Johann Gottfried, 127
Herodes, 52
Herodotus, 43, 120
Hieroglyphs, 114
Hildebrand, Hans, 29, 105, 128, 130, 180
Historical archaeology
  in Africa, 73, 79–82
  in Asia, 54, 62
  in Britain, 2
  and context, 156, 175–177, 182
    association, 168, 182; see also Ar-
      chaeology, association
    documentary, 168
    epigraphic, 168–169

Historical archaeology (*cont.*)
  and context (*cont.*)
    contrast, 171–175, 182
      chronological, 172
      form and content, 174
      spatial, 175
    correspondence, classification, 157,
      163–164
    and correlation, 164, 166–168, 182
      chronological, 167
      qualitative, 167
      quantitative, 167
      spatial, 167
  and cultural history, 126–128, 130–
    131, 134
  definition of, 105–106
  in Europe, 9–10,
  field of, 135, 143
  and identification, 162–164
  in Japan, 68–70, 72
  and Maya studies, 86
  methodology, 134, 179–183
  in Mexico, 83
  in the Middle East, 37
  in Peru, 91, 94
  professionalization of, 1, 7, 9
  in Scandinavia, 2, 183
  in the United States, 1, 82, 95–102,
    105, 108, 111, 126, 162, 164, 174,
    179–180
  professionalization of, 98
Historical linguistics, 118–119
Historical topography, 120–121, 163–164
Historicizing styles, 107–108, 110, 113,
  138, 140, 181
Historiography, 2–3, 77, 120, 126, 176,
  180–181
  Marxist, 122
Hitler, Adolf, 19
Hodder, Ian, 146
Holl, Augustin, 77
Homer, 18
  Iliad, 18, 121
Hsia, 67
von Humboldt, Alexander, 83

Iconography, 30, 34, 58, 113, 153, 157
Indian Archaeological Survey, 56, 58
Indian archaeology, 54–60, 62, 108, 111,
  117, 121–122
  colonialism, 54
  excavations, 57

Indian archaeology (*cont.*)
  landscape, 59
  processual, 59
Indo-European languages, 5–6,118–119,
  183
Indus, 58–59
Inka Empire, 91–95
Ironbridge Museum, 111

Jamestown, 97
Japanese archaeology, 68–72, 107–108,
  111, 114, 116, 122–123, 139,
  179
  excavations, 69–72
Jenne-jeno, 78
Jericho, 50
Jerusalem, 49
Johnson, Matthew, 130, 151
Jones, William, 55
Josephus, 52

Kardulias, Nick, 25
Katsura, 123
Keightley, David N., 67, 172, 174
Kemp, Barry J., 43, 146
Kenyon, Kathleen, 53
Khaldûn, Ibn, 176
  Muqaddimah, 176
Kharosthi, 56
Khipu, 94
Kilwa chronicle, 78
Kojiki, 70
Koldewey, Robert, 46
Kossinna, Gustav, 122
Krautheimer, Richard, 151, 156

Larsen, Mogens Trolle, 47
Lassus, J. B. A., 110
Latin, 147
Laurence, Ray, 46, 169
Lauriya Nandangarh, 57
Layard, Austen Henry, 44
Leake, William Martin, 14, 121, 164
Leakey, Louis, 75
Leakey, Mary, 75
von Leibniz, Gottfried Wilhelm, 62
Leone, Mark, 102, 171
Li Ji, 63
Lima, 93
Linear B, 17–18, 21, 24, 169, 171
Little, Barbara J., 134

London, 14
  British Museum, 40
  Euston Station, 111
Lu Dalin, 117
Lund, 30

Malik, S. C., 59
Mao Zedong, 62, 65–66
Marcus, Joyce, 87, 90
Mariette, Auguste, 40
Marshall, John, 58
Marxism, 140
Marxist historiography, 63, 65
Masada, 52
Material culture, 34–35, 37, 72, 78–79,
        90, 98, 101–102, 113, 116–117,
        119–120, 126–127, 132, 147,
        153, 156, 174
  and function, 161
  and text, 4, 61, 102, 113, 135, 144–145,
        167, 179, 181
    definition of, 146, 180
Materialism, 124; *see also* Archaeology,
        Marxist
Mathews, Peter, 87
Mauryan Empire, 57–58,
Maya culture, 84, 86–90
  historiography, 87
  script, 114, 116, 161
Medieval archaeology, 1, 9, 25–26, 29–34,
        108, 111, 114, 123, 126, 134, 140,
        180
  in Britain, 34
  in Denmark, 29
  in France, 31, 34
  in Germany, 29, 31, 34
  in Italy, 34
  methodology, 33
  North European, 7, 182
  in Poland, 30, 34
  profesionalization, 29–30
  research committees
    Churches Committee, 31
    Deserted Medieval Village Research
        Group, 31
    Medieval Pottery Research Group,
        31
    Moated Sites Research Group, 31
    Urban Research Committee, 31
  in Scandinavia, 34
  in Sweden, 29, 31
Megiddo, 50

Meiji Restoration, 69
Mesopotamia, 5, 47–48
Mesopotamian archaeology, 38, 43,
        46–47, 49, 107, 117, 121
  professionalization of, 45
Messenia expedition, 21
Mexican archaeology, 83, 107–108, 114,
        116, 139
  excavations, 84, 86
  processual, 84
Mexican Revolution, 84
Mexico City, 84, 86
Miletus, 17
Miller, J. Maxvell, 53
Mokkan, 72, 116, 169
Montelius, Oscar, 131, 137
Moreland, John, 35, 148
Morley, Sylanus G., 87
Morris, Craig, 95
Morris, Ian, 25, 167, 180
Mussolini, Benito, 19
Mycenaean culture, 17

Napoleon, 38
Nara, 70
  Horyuji temple, 71
Nasser, Gamal Abdel, 40
Nazism, 19
Nefertiti, 40
Netherly, Patricia J., 95
Niebuhr, Carsten, 44
Nietzsche, Friedrich, 143
Nihon-bunka-ron, 69
Nihon Shoki, 70
Nile, 41
Nilsson, Sven, 131
Nippur, 47–48
Nova Scotia, 162
  Fort Louisbourg, 162
Nubia, 78
Nuremberg, 26
  Germanisches Nationalmuseum, 26

Olduvai, 75
Olympia, 17, 137
Opium War, 63
Oracle bones, 116
Oral tradition, 5, 76, 79–81, 86, 106,
        126–127, 151, 153, 162, 164
  and writing, 6
Ottoman Empire, 10, 14, 25, 37, 44,
        49

Paddayya, K., 60, 142
Palaima, Thomas G., 169
Panofsky, Erwin, 167
Paris, 38
  Notre Dame, 110
Pausanias, 14, 17
Peloponnese, 15
Perry, W. J., 38
Peru, 2
Peruvian archaeology, 91, 179
  excavations, 94
  processual, 94
  professionalization of, 93
Philological archaeology, 117
Picasso, Pablo, 74
Piedras Negras, 88
Piggott, Stuart, 59
Pirenne, Henri, 140
de Pizzicolli, Ciriaco (Cyriac of Ancona), 114
Plymouth, 164
  Parting Ways, 164
Polanyi, Karl, 95
Pompeii, 11, 18, 169
Postgate, Nicholas, 49, 152
Potter, Parker B., 102, 171
Priene, 17
Prinsep, James, 56
Pylos, 18, 24, 169

Quadriga, 176
Quin, 65–66, 110

Radiocarbon datings, 160
Rahtz, Philip, 34, 162
Rangaku, 68
von Ranke, Leopold, 137
Ravenna, 10
Rawlinson, Henry, 44
Realia, 117–119, 181
Redin, Lars, 35
Renfrew, Colin, 119
Revett, Nicholas, 13
Ricci, Matteo, 62
Rickman, Thomas, 2, 27, 108, 136
Riegl, Alois, 19
Rigveda, 121
Robinson, Edward, 49, 121, 164
Rodwell, Warwick, 27
Roman archaeology, 24
  provincial, 9–10

Roman Empire, 58, 110
Rome, 10, 38, 117
  Ara Pacis, 19
  EUR, 19
  Instituto di Corrispondenza Archaeologica, 15
  Via dell'Impero, 19
Roosevelt, Franklin D., 97
Rosetta Stone, 38
Rumpf, Andreas, 20
Runes, 114, 169

Sachen und Wörter: see Artifacts and text
Said, Edward W., 38
Sanskrit, 55, 147
Sawyer, Peter, 2
Schliemann, Heinrich, 16–17
Schmandt-Besserat, Denise, 49
Schmidt, Peter R., 82
Schopenhauer, Arthur, 20
Schuyler, Robert L., 102, 162
Scott, Eleanor, 24
Seidenspinner, Wolfgang, 34
Semitic languages, 118
Sendero Luminoso, 94
Sepoy Mutiny, 56–57
Serlio, Sebastian, 109, 154
  L'archittetura, 109, 154
Shamanism, 67
Shang dynasty, 63–65, 67, 172, 174
Shanks, Michael, 132
Shelmerdinc, Cynthia W., 169
Sima Qian, 62
Skanör, 35
Snodgrass, Anthony, 23
South, Stanley, 101, 167
Spinoza, Baruch, 62
Stahl, Ann Brower, 80
Stephens, John Lloyd, 84
Steuer, Heiko, 158
Stockholm, 128
Stone, Elisabeth C., 47
Stoneham, MA, 133
Stuart, James, 13
Sumatra, 162–163
  Srivijaya, 162–163
Sumerian writing, 47, 147
Sung dynasty, 62–63, 67, 114, 117, 144
Susa, 171
Sutton Hoo, 162
Svenbro, Jesper, 25
Synchronic perspectives, 140–141

Tacitus, 127
Germania, 127
Tang dynasty, 65
Tawantinsuyu: *see* Inka Empire
Tello, Julio C., 93
Teotihuacán, 84
Theoderic the Great, 10
Thomas, Rosalind, 24
Thomsen, Christian Jürgensen, 2, 26, 136
Thucydides, 120
Tikal, 86, 90
Tilley, Christopher, 132
Titicaca, 94
Tiwanaku, 94
Tokugawa period, 68
Toltecs, 84
Trigger, Bruce G., 42, 75–76, 96, 130, 139
Troy, 18, 137
Tucson, 1
Tutankhamen, 40

Uhle, Max, 93
Ur, 46
Uxmal, 85

de la Vega, Garsilaso, 91
Vansina, Jan, 80
Vassunda, 114
   Ala, 114
Vázquez León, Luis, 86
Verona, 109
Vijayanagara, 61
Viollet-le-Duc, Eugène Emmanuel, 27, 110

Viru, 94
Vitruvius, 108, 161
Volkslieder, 127
Voltaire, François de, 55, 62
Vulci, 14

Wainwright, F. T., 158
Ward-Perkins, John, 21
Washington, D.C., 84
   Carnegie Institution, 84
Washington, George, 98
Welinder, Stig, 132
Wenskus, Reinhard, 35, 171–172
Wharram Percy, 125
Wheatley, Paul, 67
Wheeler, Mortimer, 58, 158
White, Hayden, 6
Wienberg, Jes, 35, 167
Willey, Gordon, 2, 94
William I, 28
Williamsburg, 99, 111
Winckelmann, Johann Joachim, 12, 19, 108, 138
Winkler, Hugo, 44

Xian, 66
Xia Nai, 67

Yadin, Yigael, 52
Yellow River, 65
Yinxu, 63–64, 67, 116
Yucatán, 84

Zanker, Paul, 20
Zhou dynasty, 65